STAFFORDSHIRE COAL MINES

Helen Harwood

In memory of Albert Cooper and Philip Cooper.

First published 2018

Amberley Publishing
The Hill, Stroud
Gloucestershire, GL5 4EP

www.amberley-books.com

Copyright © Helen Harwood, 2018

The right of Helen Harwood to be identified as the Author
of this work has been asserted in accordance with the
Copyrights, Designs and Patents Act 1988.

All rights reserved. No part of this book may be reprinted
or reproduced or utilised in any form or by any electronic,
mechanical or other means, now known or hereafter invented, including photocopying and recording,
or in any information
storage or retrieval system, without the permission in writing
from the Publishers.

British Library Cataloguing in Publication Data.
A catalogue record for this book is available from the British Library.

ISBN 978 1 4456 7787 3 (print)
ISBN 978 1 4456 7788 0 (ebook)

Origination by Amberley Publishing.
Printed in Great Britain.

Acknowledgements

I would like to thank Mel, the Museum of Cannock Chase and the National Newspaper Archives for making this book possible.

Every effort has been made to attribute copyright to the images. Any errors will be amended at the first opportunity.

Introduction

In the 1960s, a man wearing a leather back protector, his hands and face covered with coal dust, stood by the gate into the back yard and loudly shouted 'Coal!' before opening the gate and returning down the back entry to the midnight-blue flatbed lorry with the letters 'NCB' emblazoned in yellow on the cab doors. His workmate was already standing with his back to the truck, arms raised to his shoulders, ready to grab the sack of coal and carry it on his back along the entry before depositing it in the recipient's coal house. The two men would continue until the required number of sacks had been delivered. Like all the jobs directly connected with coal, it was hard and dirty work.

Coal played an important part in the lives of Staffordshire people for centuries. It was used in Britain in prehistoric times; the Bronze Age Welsh, for example, fired their funeral pyres with it. Later, when the Romans invaded, they loved its beauty, calling it gagate – which would become jet – and made jewellery with the best quality coal, while the softer coal became a popular source of fuel.

In North Staffordshire, archaeologists found evidence of coal and iron excavations from as early as the second century AD at the Romano-British settlement at Holditch when they investigated the site in 1956. There it was discovered that the site had been a centre of metalwork for both military and civilian purposes, using coal to smelt ironstone. Between AD 70 and AD 160 workshops with solid floors and plastered walls were built, linked by service roads. However, when the Romans left, the native British returned to burning wood and coal seams served as landmarks or boundaries of land.

The Old English word 'coll' is a medieval abbreviation of charcoal, while the word coal as we know it was used to describe 'pit coal', 'stone coal' or 'sea coal'.

In 1282 an early coal mine in the Manor of Tunstall was valued at 14s 8d (74p). Later, in 1293, a 'manufactory of iron' was recorded at Madeley and coal was excavated in the Manor of Madeley in 1333. In 1308, Keele Manor recorded coal and iron mining. Then there was coal mining at Norton-in-the-Moors in 1316 and at Lee Fields in the fourteenth century on land belonging to the Cistercian monks of Hulton Abbey.

Crown administration documents in the sixteenth century list a coal mine owned by Hulton Abbey and leased to Thomas Fox in the 'Field of Hulton', or 'Lee Fields', which were probably west of the River Trent and in the region of what was later to became known as Foxley. Additionally, c. 1527 the abbey owned a smithy at Horton Hay and a coal mine in Hanley. In the medieval period coal was used mainly for manufacturing

Introduction

The remains of Hulton Abbey, where coal was mined in the sixteenth century. The ruins are now grassed over to protect them. (Author's collection)

as opposed to domestic purposes. It is unlikely that Hulton Abbey's mine provided the abbey with coal as, apart from the warming room, it was unheated. Presumably the monks would have used some coal for their kilns; however, accounts from the Court of Augmentations record that the rent from the abbey's mine was used to fund their hospice, which provided guests with hospitality.

During the reign of King Edward I (1272–1307) wood became expensive for the growing population and under Forest Law penalties for taking it were severe. As a result, easily accessible coal became a necessity for the poor. But in a time of open hearths and no chimneys, smoke would make its own way out through louvres in gable ends and gaps in the roof, making the atmosphere overwhelming. In 1306 King Edward I banned coal burning under severe penalty of a fine (though some were also tortured) as the sulphurous smoke was making people ill. His own mother, Eleanor of Provence, became so sick from the polluted air around Nottingham Castle that she left the town. Generally the law was ignored, even when his successor, King Edward II, made it a capital offence to burn coal in London, and one man was actually executed for doing so. Later, King Richard II, Henry V and Queen Elizabeth I all tried and failed to prevent coal burning.

The township of Quarnfield in the Staffordshire Moorlands, which includes the village of Flash, had coal workings at Golditch Moss in 1564 and Black Clough in 1602. Peter Higson rented a house at Golditch Moss with 37 acres of land, including coal workings at Quarnfield Moss, in 1634 for a lease of twenty-one years at £12, plus two fat hens per annum! In 1673 coal was dug from outcrops along the stream and then, later, William Wardle of Boosley Grange in Fallfield Head leased mines in Alstonfield parish in 1677 and in his will dated 1718 he requested that Sir John Harpur, Lord of Alstonfield, extend the lease to benefit William, his son. In 1765 Henry Harpur rented mines at Golditch and Knotbury to George Goodwin and John Wheeldon of nearby Derbyshire and James Slack of Knotbury for twenty-one years at £10 15s (£10.75) per annum. Possibly Goodwin and Wheeldon payed the rent and Slack managed the mine.

In 1750 Ralph Leigh of Burslem collected coal from Whitfield and landowners – for example, Bowyers, Heathcotes and Sneyd – began to realise the increasing value of the mineral deposits under their respective estates. 1775 saw iron ore and coal hewn from shallow workings in Apedale, while two years later, in 1777, James Brindley discovered a significant coal seam while building the Harecastle canal tunnel, which gave landowners greater encouragement to excavate their estates.

From the late eighteenth century, the Upper Silverdale valleys and Podmore Hall were also extensively mined, with the earliest workings in the upper valleys and on the ridges in between where there were fewer drainage problems. To begin with coal was surface mined but later, in the nineteenth century, the workings gradually deepened. In 1833 Apedale was at a depth of 2,500 feet; however, the shaft was only 720 feet deep, and then an 800-foot inclined plane, driven by a steam engine, drew the coal to the bottom of the shaft – a mechanical wonder of innovation.

The entrance to James Brindley's Harecastle Tunnel.

Introduction

Following the end of the Napoleonic Wars in 1815, England saw an industrial expansion. At Leycett, workings carried the names Nelson, Victory and Blucher.

In the township of Tunstall to the north, coal was regarded as of good quality for the pottery kilns. In 1843 Little Pits was owned by H. H. Williamson, Clanway by Messrs Child & Clive, Greenfield by Messrs W. & E. Adams and Botany Bay was owned by Joseph Heath & Co., while Oldcot was owned by Robert Williamson, who also ran mines on his neighbour's estates along with his own. The coal was partly used for the pottery kilns in Tunstall, but the majority was sold on the general market, transported by railway to an underground branch canal, which later joined the Trent & Mersey Canal in the old Harecastle Tunnel.

In 1816 a swimming baths opened that was heated with water supplied from the neighbouring engines of Bycars (or Bycrofts) Colliery, but only a few years later this fell into disuse and was closed.

In 1843, on the Fenton Park Estate, owned by Sir Thomas Boughey and Mr Armistead, there stood a large colliery run by W. T. Copeland and others as the Fenton Park Company. Meanwhile, in Longton, Lane End and nearby Fenton, the mines were run by William Hanbury Sparrow for the Duke of Sutherland, Charles Smith Esq., the Fenton Park Co., the Oldfield Colliery Co., Thomas Wynne & Co., Mossfield, Mr Ralph Handley and others. The deepest shafts at the time were 320 yards down – far deeper than those generally sunk in the area.

The introduction of hot-blast furnaces within the iron industry in the years 1840–70 saw the demand for coal increase, where previously the majority had been used for kilns in the pottery trade. It was recorded in 1862 that North Staffordshire's workable coal seams extended for 147 feet – 25 feet more than anywhere else in Britain – producing 3 million tons of coal in 1864. By the end of the nineteenth century there were few coal fields able to supply such a variety of coal to suit the pottery industry, iron foundries and the steam engines of the railway. At this time coal mining was growing rapidly to feed the Industrial Revolution and finished any hopes of those wishing for its prohibition. Indeed, to be against coal was to be against progress and employment. Coal usage grew by 100 per cent between 1800 and 1900. Moreover, in 1939 North Staffs produced 7.2 million tons – a total greater than the Black Country and Cannock Chase together.

In the South Staffordshire Coalfield – which for historical reasons includes the areas that found themselves in the West Midlands region after the 1974 boundary changes – documents record mining in the thirteenth century in Rushall and Walsall, as well as at Sedgley in 1273 and Kingswinford in 1291, while a mine at Cannock in 1297 was valued at 48*s* (£2.40) a year. In the next century, in Wednesbury in 1315 and in Longdon and Cannock Chase in 1305 five separate mines that were leased at 6*d* (2.5p) per pick were recorded as being owned by the Bishop of Lichfield. Later, in 1497, three men held the lease of a coal mine in Beaudesert Park. Following the English Reformation of the 1530s, the Bishop's land, covering a large area of southern Cannock Chase, fell into the hands of Sir William Paget of Beaudesert Hall, who, along with the Lords Dudley, Enville and Stafford, began mining on a greater scale. Later still, coal from Beaudesert Park was sent to Tutbury Castle when Mary, Queen of Scots was imprisoned there on several occasions from 4 February 1569 until leaving for the last time on 24 December 1585. The estate was confiscated by Queen Elizabeth I after it was discovered that Thomas and Charles Paget were implicated in the Throckmorton Plot of 1583, which was a plan to murder the

Above: Beaudesert Hall.

Left: Tutbury Castle, where Mary, Queen of Scots was imprisoned.

Introduction

Queen and replace her with Mary, Queen of Scots. They fled to France and it was another twenty years before Thomas's heir could reclaim the estate, during which time Gilbert Wakering of Bloxwich, who also held the lordship of Wombourne, held the rights to mine the land. Apart from this period, the Paget family, unlike other estate owners who leased out workings, preferred to oversee their mines themselves.

In the seventeenth century coal was worked in the area, mainly south of Watling Street, now the A5, at Cheslyn Hay, Easington and Great Wyrley, initially by bell pits and then shallow shaft mines. Later the development of steam engines to improve ventilation and pump out water meant shafts could be sunk to a much deeper level. The first large-scale mine was Heath Pit, begun in 1833 by Lord Dartmouth of Sandwell Hall, reaching a depth of over 900 feet.

By the early nineteenth century there was serious competition between the Paget family, the Marquis of Anglesey of Beaudesert and the Chetwyn-Talbots, the future Earls of Shrewsbury of Ingestre Hall, as they both extended their mining interests; this was until 1854, when Earl Talbot leased the Marquis's mines and the local pits had a single management for the first time. In 1835 there were fifty operational mines and by 1856 1.29 million tons of coal was produced, increasing to 3.9 million tons in 1870 to feed the Black Country's iron industry boom. In 1861 the population of the Urban District of Cannock was 2,913; by 1891 it had grown to 20,613 following the economic upturn after the depression of the 1880s. This was partly as a result of older mines in the south being abandoned in favour of those in the north-west, wherein the thirty-three pits on Cannock Chase were producing as much coal as the 276 old mines had done in the main South Staffordshire Coalfield.

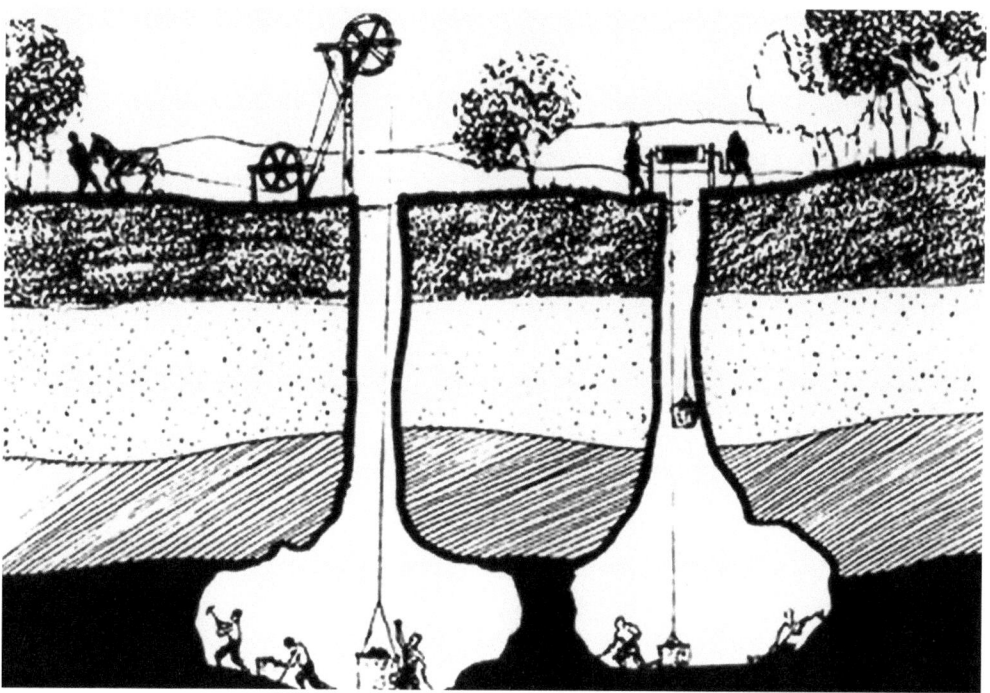

Diagram of an early bell pit.

Ingestre Hall, home of the Chetwyn-Talbots – the future Earls of Shrewsbury. (Author's collection)

The North Staffordshire Coalfield

Apedale

Apedale Valley was possibly the most productive valley in the country during the Industrial Revolution. Producing coal, ironstone and clay, it empowered other neighbouring manufacturers to introduce mass production. A major local landowner, Sir Nigel Gresley (1726–87), began work in 1775 on the construction of one of the area's earliest canals. It would allow him to transport ironstone together with coal from his pits in Apedale to the centre of Newcastle-under-Lyme to feed an established iron-working industry, which particularly benefitted the manufacturing of nails – hence the 'Ironmarket'.

By the mid-1830s the North Staffordshire Railway Company was formed and in the 1860s they built a branch line to Apedale with a junction on the Newcastle to Audley line, making it easier and more economical to transport coal and iron by rail. As the small, shallow mines became exhausted, larger, deeper pits were opened, and in the 1890s Apedale had three main collieries – Watermills, Burley and Sladderhill – mining a total of nine different coal seams. Burley had the thickest coal seams; however, to reach them the shafts were 475 yards (434 metres) deep and were apparently the deepest in the country at the time. Meanwhile, Watermills had one shaft to a depth of 350 yards (or 320 metres), while Sladderhill Pit was comparatively shallow, at 200 yards (183 metres) deep. The latter received accumulated water from the other mines and so housed the pumping engines. Apedale also had two further mines: the New Recovery Pit and the Lilly Pit. Both were 175 yards (or 160 metres) deep, but their coal and ironstone were of a poorer quality than that of their larger neighbours.

In 1890 the Midland Coal, Coke & Iron Company took over the running of Apedale's industries, along with the mines at Podmore Hall and Halmerend.

Difficult economic conditions arose following the end of the First World War in 1918 and Burley Pit, the last working mine in Apedale Valley, closed in 1926.

Mining was to return again, however, when a new drift mine was opened in 1941 by Holditch Colliery, known as the Apedale Footrail Colliery. This in turn closed in 1969

Above: Apedale mining museum, on the site of the Apedale Colliery. (Author's collection)

Left: The pithead wheel and coal tub monument at Apedale Country Park. (Author's collection)

The coal tub at Apedale Country Park. (Author's collection)

Burley Pit at Apedale Colliery.

Coal picking at Apedale c. 1926.

as it was considered to be uneconomical. This was a controversial decision as the Aurora Mining Company reopened it later and were working seven drift mines in the valley by the 1980s before they ceased trading in 1998. The site is now home to the Apedale Heritage Centre.

Berry Hill

As early as 1628, Robert Bagnall and Longton mine owner Jeremiah Smith worked the coal seams at Berry Hill, Fenton, and in the 1840s W. T. Copeland began mining coal and ironstone. By the 1860s W. Bowers – who was prominent in the Cheadle Coal Field – had purchased the site. By 1870 he'd built an ironworks on the site, which was to remain until the early twentieth century. In 1872, however, an underground explosion killed six men. They died from severe burns when a naked flame ignited gas. At the inquest, the Coroner exonerated the management and blamed the men for using a naked flame underground. Later, the bereaved would probably have followed the North Staffordshire custom of leaving a miner's clogs on his grave when he died in the pit.

Following Bowers' death, the site was purchased by Henry Worthington in the 1880s, which by now comprised almost 100 acres with mineral rights on a further 500 to 700 acres, and where a brickworks was built that continued into the 1960s. Around the beginning of the First World War (1914–18) Worthington's company went into receivership and was subsequently bought by John Slater, who in 1919 set up John Slater

Berryhill Colliery.

Limited to manage Berry Hill and New Haden (Cheadle) Collieries. He then went on to acquire an interest in a holding company, of which Berry Hill and New Haden became associated, trading as Berry Hill Collieries Limited, while also purchasing a majority share in the holding company that operated Chatterley Whitfield Colliery. However, due to financial pressure Slater's holding company went into receivership, though he was still found to be in control of some companies, including Berry Hill Collieries Limited. By the early 1930s, Slater, ever the entrepreneur, became a board member of Settle Speakman, & Company, which coincided with a new No. 1 shaft sunk to a depth of 750 yards at Berryhill. Coal raising was later stopped on the site and its allotted tonnage was transferred to New Haden before returning to Berry Hill in 1933, but from then on all coal winding was concentrated on the new shaft.

Due to geological conditions, the pit's output gradually declined and the mine closed in 1960. Later, in 1963, the old pithead baths became a mines rescue station, while in July 1966 the National Coal Board opened their area headquarters on the site. It was also home to the Area General Road Transport Garage, which was later sold to M. Nield, who subsequently left the site in 1993. The site is now Berryhill Industrial Estate.

Birchenwood
Birchenwood Colliery, Kidsgrove, lay close to the Potteries Loop Line and Harecastle Tunnel, where coal had been discovered when originally excavated. Indeed, Thomas Gilbert,

the Duke of Bridgewater's representative, pioneered large-scale mining in the area. Birchenwood Colliery began around 1891 with five shafts and a footrail for winding, ventilation and pumping, while the only coal winding shaft was No. 18 – the main shaft. Virtually all the coal produced was used for coke and other byproducts. Following landowner Robert Heath's death in 1893, the business passed to his two sons. Meanwhile, the company operating the Birchenwood iron and steelworks closed, allowing the Heath brothers to purchase the entire site and found the Birchenwood Colliery Company. However, by 1925 the colliery was in liquidation and the company proposed to operate as a working co-operative. The employees rejected the offer, resulting in the Kidsgrove Collieries Company being founded in June that year, with the Heath family as directors.

Around 1 p.m. on Friday 18 December 1925, Assistant Surveyor G. H. Forrester discovered an unusual smell in the return air. Overman George Wilcox went to check an hour later and found nothing unusual, for which Forrester apologised. Then, close to the end of the shift, an explosion at 4.15 p.m. in the 7-foot Banbury seam killed seven miners and seriously injured another fourteen. Of the fifteen men in the upper panel, only eight survived. At a depth of 1,100 yards and some three quarters of a mile from the shaft, it took several days to recover the bodies, many of which were badly burned. A verdict of accidental death was recorded.

Later, the colliery passed to another company, run by Mr James Cadman, before closing around 1932. However, coke and other byproducts continued to be produced using coal from the Biddulph mines. The last coke to be made was in May 1973.

Chatterley Whitfield

In an area rich with coal seams, Chatterley Whitfield Colliery opened in 1838. In 1853 Hugh Henshall Williamson of Greenway Bank Hall also mined there. A year later the coal masters compelled the North Staffordshire Railway to build the Biddulph Valley Branch line or they would build it themselves. Work began in 1858 and when completed in 1860 the railway passed half a mile from Whitfield Colliery. In November 1867, a month before Hugh Henshall Williamson's death, 'Gentlemen of Tunstall' took ownership of the mine as the Whitfield Colliery Company. Then, in 1872, Mr Charles Homer, the Managing Director of Chatterley Iron Company, purchased the mine on behalf of his company.

At 3.15 a.m. on 7 February 1881, due to the misuse of an underground blacksmith's furnace, a fire and explosion killed twenty-four men and boys, for which the Manager, Mr Thompson, was tried and acquitted of manslaughter at Stafford assizes. Following the explosion, Middle Pit shaft was deepened and a new shaft was completed in 1883. When the company fell into financial difficulties a year later, it fell to three liquidators – one of whom, John Renshaw Wain, was the previous company secretary. It was to be his son, Edward Brownfield Wain, who was to take Chatterley Whitfield into its 'Golden Age' in the 1890s.

Coal stopped being drawn at Institute shaft in 1955 and Middle Pit shaft in 1968, with the last of the coal being lifted in 1976. Chatterley Whitfield closed on Friday 16 January 1976. Today it is the biggest deep mine site in England and a National Monument.

Original NCB Western Area sign outside Chatterley Whitfield Colliery. (Author's collection)

The pithead winding gear at Chatterley Whitfield. (Author's collection)

Coal waggons at Chatterley Whitfield. (Author's collection)

Diglake

Diglake Colliery, Bignall End, was situated a short distance from Audley railway station, where the line crossed the Chesterton Railway. It was owned by Sir Thomas Boughly and was operated by William Rigby & Sons with a connection to the Keele–Alsager branch of the North Staffordshire Railway. There were three shafts, the third of which was some 400 yards from Diglake No. 1, which it had an underground connection to. It had previously been part of the closed Hall Pit and was re-opened as a winding shaft.

At 11.30 a.m. on Monday 14 January 1895, Fireman William Sproston fired a shot, the result of which caused a torrent of water from the old Rookery Pit workings to flood the lower part of the mine. That winter had been particularly hard, with heavy snowfalls accumulating into 8-foot drifts; rain and a partial thaw culminated in large

The Diglake memorial, dedicated to the seventy-five miners who died on 14 January 1895.

amounts of water, which added to the flood. Seventy-eight men and boys – some as young as thirteen – died. Those who survived had Messrs Bateman, Boulton, Watts, Dodd, Carter and Hinkley to thank for their lives. The search for survivors continued over the following week to no avail, though on the next Saturday rescuers brought ten pit ponies, along with a cat and two kittens, to the surface.

Following the accident all mining activities were undertaken from a new Rookery Pit, which was built several hundred yards to the north-west. Thirty-seven years later, on 12 August 1932, a heading reached the abandoned Diglake mine; on 3 September a skeleton was discovered, unfortunately without any identification, followed by another on 6 March. A decision was taken not to proceed and the workings were sealed, leaving the remaining seventy-two miners to rest where they had died.

Florence

Florence was named after the eldest daughter of the 3rd Duke of Sutherland, who initially worked the pit privately. Work on the sinking of two shafts began in 1874 before the Florence Colliery Limited Company was formed in 1891, followed later by the Florence Coal & Iron Company in 1896. A third shaft was sunk in 1916. Following nationalisation in 1947, a £7 million modernisation programme began, culminating in a new coal plant, horizon mining and electric winders. There had been underground connections with Hem Heath and Parkhall over the years, and in 1974 Florence was merged with Hem Heath and became part of the Trentham Superpit. It finally closed in 1994.

Florence Colliery, Longton.

A coal wagon from Florence Colliery. (Author's collection)

Trentham Hall, the former home of the Duke of Sutherland, who initially owned Florence Colliery.

Foxfield

Foxfield Colliery at Dilhorne, originally known as the 'Manns' Pit', was the last deep mine of the Cheadle Coalfield, with coal digging in the Dilhorne, Godley Brook, area dating back to the seventeenth century. Indeed, Samuel Bamford mined coal on what was to become the site of Foxfield Colliery. However, it was in 1880 that John and Enoch Mann of Blakeley House, Cresswell Ford, began work to sink the first shaft to a depth of 735 feet, and although coal was being mined in 1884, it was not until 1888 that the shaft was completed. There then followed a second shaft dug to a depth of 1,000 feet, which remained in use throughout the pit's productive life.

Foxfield Colliery. (Author's collection)

By 1893, when Messrs Mann had registered the Foxfield Colliery Company Limited as a private limited company, it included a private railway, and the 3½-mile line joined the North Staffordshire Railway at Blythe Bridge. However, due to the refusal of a nearby landowner to permit the building of the line over his land, the company had little choice but to take the line around his boundary and through Dilhorne Wood, entering the pit in the opposite direction to Blythe Bridge with a gradient of 1 in 19. The company had intended to connect the mine to the proposed Cheadle Railway, but it was 1901 before this was built.

On 23 August 1927, the Foxfield Company went into voluntary liquidation and was bought, along with all freehold, leasehold and mining rights, by Parkhall Colliery Company Limited of Cheadle. Meanwhile, the final meeting of the old company took place on 3 October, 1927. Both Parkhall and Foxfield collieries continued to be run independently, despite there being an underground connection, though Parkhall later became a pumping pit. In the 1930s Foxfield underwent extensive modernisation, with the construction of new coal screens and concrete headgear, which was unique to the Cheadle area. There was further expansion following nationalisation in 1947, despite Cash Heath Wharf closing in 1949; pithead baths were built in July 1949 and a new spoil tip was added around 1950. Foxfield, however, was never profitable due

The two sets of winding gear at Foxfield. (Author's collection)

The shaft head at Foxfield. (Author's collection)

to its limited output and high costs. It closed in October 1965, having exhausted any remaining economic reserves. Tean Minerals went on to purchase the site, followed by Fergusson Wilde. Meanwhile, the branch line to Blythe Bridge was bought by the Foxfield Railway Society, who, together with a local group of enthusiasts, tried to develop a mining museum in the remaining colliery buildings in 1995, but the plans were unsuccessful.

Glebe

Glebe Colliery, Fenton, named after its association with the Glebe lands of Stoke's St Peter Ad Vincula Church, was opened in 1865 by John Challinor & Company on a site close to both industrial and heavily populated areas. Therefore, a great shaft pillar was deemed necessary to ensure the safety of the surrounding buildings. However, this meant that most of the workings were some distance from the shaft. A year later, at 3 a.m. on Wednesday 18 July 1866, an explosion killed four miners and injured two more. Naked candles were said to be the cause.

When Glebe was passed into the ownership of John Heath & Company at the turn of the century, operations were growing. However, following nationalisation in 1947, another explosion on 13 June 1963 took the lives of three miners in a colliery that had been losing money for some time. Glebe closed in spring 1964, with the pithead gear

Left: The coal tub monument to Glebe Colliery on the pit's landscaped spoil heap at Fenton. (Author's collection)

Below: The plaque on the coal tub monument. (Author's collection)

CITY OF STOKE-ON-TRENT
GLEBEDALE PARK RECLAMATION SCHEME
(FENTON TIP)
Official Opening by
MR. DEREK EZRA
Chairman of the National Coal Board
on Wednesday, 2nd May, 1973
Councillor William Austin, Lord Mayor
Councillor J. Monks-Neil
Chairman, Land Reclamation Committee
Councillor J. Westwood
Vice-Chairman, Land Reclamation Committee

L. Keith Robinson, LL.B.,
Town Clerk

J. W. Plant, F.R.I.B.A., F.R.T.P.I.
City Architect, Planning and
Reconstruction Officer

S. N. Mustow, B.Sc., C.Eng.,
City Engineer & Surveyor

P. Dyer, N.D.H., F.Inst. P.R.A. (Dip)
Director of Parks, Recreation and
Cemeteries

being taken down and the shafts filled in by the following August. Today the dirt tip has been reclaimed and is now Glebedale Park, which was opened by Derek Ezra, Chairman of the then National Coal Board, on Wednesday 2 May 1973.

Hanley Deep

Hanley Deep was so named because at one time it was the deepest mine in North Staffordshire. It was owned by the Shelton Coal & Iron Company Limited under the 2nd Earl Granville, the Right Honourable Granville George Leveson-Gower.

The pit's early shafts were sunk to a depth of 500 yards in 1854 and these were later widened and deepened to a depth of 880 yards. Apparently, the excavated earth was taken to Northwood graveyard, where it added a further 6 feet of covering. As plague victims were buried there, it was uncertain whether any remaining infection could be released if the bodies were ever disturbed, so layering an extra 6 feet of soil gave some insurance against that happening.

On Monday 21 November 1881, a work party descended the mine to re-set a damper; however, before they could begin work at 5 a.m., an explosion ripped through the pit, killing two people and injuring a further seven – twenty-five years after a similar explosion had occurred at Hanley Deep.

The company was bought by John Summers of Shotton, North Wales, in November 1920 and later, in the 1950s, with economic coal reserves gradually being depleted, an underground connection was made with Wolstanton Colliery. Production finished at Hanley Deep in July 1962 and the shafts were filled in soon afterwards, although the last winding engine survived until 1994. The site is now Hanley Forest Park.

Hanley Deep Colliery.

The winding wheel memorial to Hanley Deep, which is now Hanley Forest Park. (Author's collection)

Hem Heath

The Duke of Sutherland cut the first turf of the new Hem Heath Colliery on 30 July 1924 for the Stafford Coal & Iron Company Limited. Initially there had been concern as to the effect a new pit would have on Trentham village, so to compromise an agreement was reached to process coal at Stafford Colliery, to the north. Therefore, only No. 1 shaft and its accompanying buildings were to occupy the site, with a second access being made through an underground connection with Kemball Pit. However, there was a considerable delay in sinking the shaft due to the discovery of a previously unknown fast-flowing stream, and the ground had to be frozen before work could recommence.

During the Second World War a number of 'Bevin Boys' worked at Hem Heath, one of whom went on to become a Cistercian monk. Following nationalisation, in 1947 plans were developed for No. 2 shaft to be sunk to 4,000 feet, 24 feet in diameter (the third

Hem Heath Colliery (1924–50).

deepest in the country), with Sir Ben Smith, Chairman of the West Midlands Division of the National Coal Board, making the first cut on 25 January 1950. With a second shaft the underground access from Kemball was no longer required and that tunnel was sealed. Meanwhile, No. 1 shaft was deepened.

In August 1956 No. 2 overtook the older No. 1 to become the main coal-drawing shaft. Then, in the early 1960s, Stafford Pit's coal preparation facilities were concentrated at Hem Heath, while in the early 1970s Florence merged with Hem Heath to form the Trentham Superpit. Mr D. J. Ezra, Chairman of the National Coal Board, cut the first turf on Friday 24 May 1974. The pits were connected underground with a 1 in 4 inclined drift from Hem Heath to Florence and a new coal preparation plant was built at the Hem Heath railhead. Although all the coal was lifted at Hem Heath, both collieries retained their separate management until the late 1980s, when they became known as Trentham Colliery (Hem Heath was Trentham West and Florence was Trentham East). The majority of coal was sold to the Midlands Electricity Board.

Coal was last produced in October 1992 before the pit closed in October 1994. It then reopened when Coal Investments plc took out a fifteen-year lease from privately owned UK Coal and invested £5.5 million. Two years later the company collapsed with huge debts and Hem Heath finally closed in 1996. The pit's classic 'A' frame headgear was demolished in 1997 and the site is now home to housing and an industrial estate.

The Hem Heath pub, which now stands on the site of the colliery. (Author's collection)

Holditch

Holditch Colliery – also known as Brymbo – opened in 1912 when a 2,000-foot shaft was sunk by the Brymbo Steel Company of Wrexham for ironstone. Another shaft of the same depth followed in 1916 and the mine produced both coal and ironstone. In the late 1920s John Summers & Sons of Hawarden Bridge Steel Works, Shotton, who were also the owners of Shelton Steel Works at Etruria, became interested in Holditch as a source of iron and coal to supply the Shelton blast furnaces. On 11 January 1930 a new company, Holditch Mines Limited, was formed, with the output going two thirds to Shelton and one third to Brymbo, though the latter went into liquidation a year later, in June 1931. The Shelton Company bought Brymbo's share on 5 February 1932, becoming the sole owners of the mine.

At 5.45 a.m. on 2 July 1937, two coal cutters spotted flames to the rear of a coal cutting machine, which quickly developed into a wall of fire some 7 or 8 yards long across the coal face. The two men fled and were fortunate to survive; however, the fire went on to cause eight explosions, killing thirty men and injuring a further eight. There were eleven more deaths between 1949 and 1967, mostly from roof falls.

Holditch was modernised in 1947; however, the pit was still somewhat old-fashioned in that steam power continued to be used in some form until 1980. From July 1967 it

The Holditch Colliery gates, now preserved at the Apedale mining museum. (Author's collection)

The plaque on Holditch colliery gates. (Author's collection)

A coal train on the line to the Holditch and Silverdale pits.

sent 1.25 million therms of gas per year to the nearby brickworks, which fired 500,000 bricks a week, together with an input to the North Staffordshire Gas Grid. Despite heavy investment in the 1960s and '70s, the colliery closed in 1990. Today the site is a business park.

Ivy House

Ivy House and Northwood collieries in the borough of Hanley lay around 200 yards from the Cauldon Canal, and about 250 yards from Bucknall station on the Biddulph branch line of the North Staffordshire Railway. Originally the two pits were run under separate owners; however, records show that Richard and William Jennings were proprietors in 1820, followed in 1869 by the Ivy House & Northwood Mining Company. In 1874 both mines were united under a new company, run by Mr R. H. Wynne, over an area of some 200 acres. Two new shafts with a diameter of 14 feet 6 inches were sunk to a depth of around 280 yards to access the Mossfield coal, and these were worked by two engines. The site now had four shafts and a branch line to the nearby canal and railway. Five years later, in 1889, an explosion killed three miners, who were buried in Bucknall churchyard.

Jamage

Jamage Colliery, Talke, was opened by Robert Rigby in the early part of the nineteenth century and was connected underground to an adjacent pit, both of which were owned by representatives of the late John Wedgwood. The mines had two shafts – one 10 feet in diameter and the other 13 feet – and both were around 170 yards deep. On Christmas Eve 1874 an explosion at the adjacent pit resulted in seventeen deaths. Thirteen months later disaster struck again; at 4.30 p.m. on 5 January 1876 the pit suffered an explosion in the 7-foot seam, which killed five men.

In the 1870s an iron blast furnace was built, and in the 1890s a coking station, slack works and chemical works were added to increase profits from the coal byproducts.

Jamage suffered another explosion at 9.50 a.m. on Saturday 25 November 1911 when gas exploded, causing the deaths of six men by carbon monoxide poisoning and injuring fourteen others, four of whom suffered burns. It closed in 1928, though some of the mine's railway lines survived until 1975, when they fell victim to the A500 road development. The site is now occupied by the Freeport shopping mall.

Jamage Colliery.

The modern-day site of Jamage Colliery, which is now a shopping centre. (Author's collection)

Kemball

Work began on Kemball Colliery (known as the 'Duke's Pit') at Heron Cross in 1873 and was completed in 1876. It was owned by the Stafford Coal & Iron Company Limited, with the Duke of Sutherland and C. J. Homer as the major shareholders (the latter having come from Chatterley Whitfield), and was part of the Great Fenton Collieries. Earlier, the company had bought the Great Fenton Estate from John Bourne, which also included the mineral rights, and subsequently built a brickwork in 1874 and two blast furnaces in 1876. The furnaces and by-products plant closed in 1931, while the brickworks survived until the 1960s.

Later, the two shafts (known as 'Homer' and 'Sutherland') were deepened and in 1936 Cecil Speakman went on to take financial control of the company. Afterwards, in the 1940s Kemball became a training pit, where many Bevin Boys had their first underground experience.

At the time of nationalisation in 1947, the colliery's leasehold estate comprised 7,000 acres, which was worked from Hem Heath's No. 1 shaft along with the two shafts at Kemball. The pit also had a wharf on the Trent & Mersey Canal, and in the late 1950s the canal bank collapsed, with a number of workers both on a barge and the bank fortunate to escape fatal injury. With the development and separation of Hem Heath

Above: Kemball Pit.

Right: Certificate of training for employment below ground, issued from the Kemball training centre. (Author's collection)

Colliery in the 1950s, together with geological problems, Kemball was gradually run down and in the 1960s a decision was taken to transfer all coal preparation to the larger and more modern Hem Heath, along with all locomotive work, which was operated for both collieries from a purpose-built engine shed at Hem Heath Pit. Geologically, Kemball had never been an easy colliery to work and it was closed in 1968. The site went on to become the area's Surface Training Centre until 1989, when new facilities at Hem Heath replaced it. Most of the site was cleared to facilitate the A50 road improvements and a hotel now stands on the pit site. Interestingly, the 'Homer' and 'Sutherland' shafts had the last two vertical steam winding engines in North Staffordshire.

Madeley

The village of Leycett was built for coal and iron. The colliery had four shafts, which led into what was possibly two pits, and in 1839 the colliery was connected via the Madeley Branch Railway (one of the first privately owned industrial railways) to the Crewe–Stafford line close to the site of the former Madeley railway station. In 1875 it was owned by the Crewe Coal & Iron Company and the colliery was known locally as 'The Fair Lady', though where the name came from is unclear.

A series of early unexplained explosions blighted the pit. In 1871 forty-four men died, followed by another forty-one in 1879 and in 1880 a further sixty-two lives were lost. Ownership of the colliery changed hands several times, from the Silverdale Colliery Company to the Madeley Coal & Iron Company, under whose management another six men died and a further three were injured on Sunday 21 October 1883. Later came

Madeley Colliery at Leycett.

the Madeley Coal & Brick (1905) Company, followed by Madeley Collieries Limited, before nationalisation.

Due to difficult geological conditions, the mine was never a financial success and closed in 1959. The majority of the village was later demolished.

Minnie Pit

The Minnie Pit at Halmerend began in 1881 and was named after Minnie Craig, the daughter of Mr W. Y. Craig, one of the owners of the company, who lived at Alsager Hall. Work began in 1883 – with Minnie herself digging the first turf – to sink what was to become an addition to the Podmore Hall Colliery; its completion was celebrated with a supper for 108 men and officials on Friday 30 October at the Halmerend National schoolroom. The shaft was originally 18 feet in diameter. However, work was delayed when a quantity of marl together with 30 yards of extremely wet quicksand was encountered. It took six months to install an inner ring (a caisson or steam shell) 16 feet 3 inches in diameter to reach through the quicksand and sink the shaft to around 359 yards. Next, a connection with an existing shaft for ventilation at Podmore Hall Colliery was dug through unknown terrain, where two new coal seams, each 8 feet thick, were subsequently discovered. The mine headgear was built of brick outside retaining walls lined with masonry and filled with rock and bass on a concrete base. The headgear, made of pitch pine timber, weighed 70 tons and stood 55 feet high. The shaft became the Downcast shaft for the Podmore Hall Colliery, itself part of a larger company that owned the Burley Pit, the main winding pit for the Podmore Hall estate, close to Apedale, along with an ironworks, forge and coking ovens at Apedale. In 1890 the Midland, Coal, Coke & Iron Company was founded to oversee the valley's industrial operations, building its own narrow gauge line, the Apedale & Podmore Hall Railway, in order to give their mines a connection with the North Staffordshire Railway's Audley line.

The Minnie Pit, however, was to have a history of explosions due to its susceptibility to firedamp. The first, occuring on Sunday 6 February 1898, killed all the pit ponies. Then, on Sunday 17 January 1915, twenty-seven men and boys from the engineering company were undertaking repair work when an explosion at 5.30 p.m. killed nine of them, including the Colliery Engineer, Mr John White. Another victim was just sixteen years old. No evidence was found as to the cause.

Two years later, on Saturday 12 January 1918, 248 miners were underground when the Bullhurst and Banbury seams suffered a catastrophic explosion, resulting in roof falls and poisonous gas, which killed 155 men and boys – forty-eight of whom were under seventeen; indeed, many were just fourteen years old, and the eldest victim was sixty-five. Then, sadly, Hugh Doorbar, Captain of the Birchenwood Colliery No. 1 rescue team, was killed in the attempted rescue operation, bringing the final death toll to 156. Many areas of the mine had collapsed and methane gas filled the workings, making it extremely dangerous for the rescue teams. The Primitive Methodist Chapel at Halmerend became a temporary mortuary, but it was to take twelve months before all the bodies were recovered. This was the worst mining disaster in North Staffordshire.

The Minnie Pit closed on Saturday 26 April 1930 along with the all the other operations owned by the Midland, Coal, Coke & Iron Company.

Left: A monument on the Minnie Pit site dedicated to the victims of the 1918 disaster. (Author's collection)

Below: The coal tub memorial at the former Minnie Pit, which was unveiled on 15 October 1988. (Author's collection)

The North Staffordshire Coalfield

In Memory
of the 155 miners and 1 rescuer
who lost their lives in the
Minnie Pit Disaster
January 12th 1918.
Also all who lost their lives
in the extraction of coal
from this Mine.

Above: The plaque on the coal tub memorial to the Minnie Pit disaster. (Author's collection)

Right: The brass plaque on Halmerend's Primitive Methodist Chapel, bearing the names of those who died in the Minnie Pit disaster. (Author's collection)

Above: The Primitive Methodist Chapel, Halmerend, which became a temporary mortuary following the 1918 disaster. The memorial plaque to the deceased miners is on the left. (Author's collection)

Left: A poster in memory of the Minnie Pit explosion.

Mossfield

Mossfield Colliery was sunk in 1819 and had two shafts 440 yards deep with tandem headgear – both shafts lay to one side of the winding engine. Known locally as 'Owd Sal', it was designed to work the lower depths, being some 20 yards deeper than the Cockshead seam.

Meanwhile, an underground connection was made with the nearby Adderley Green Colliery (1799–1939) and its No. 9 shaft continued as a ventilation and second egress for Mossfield.

In 1830 the owners were listed as Heathcote & Woley and in 1843 as Thomas Wynn & Co. The owner was then a Mr Ralph Handley, while in 1875 and 1888 the owners were listed as Hawley & Co.

The Cockshead seam was noted for gob fires. 12 September 1889 saw the beginning of a number of explosions and on 14 October there was a characteristic smell noted, the reports of which were disregarded. Two days later, on 16 October 1889, a massive explosion ignited by a spontaneous gob fire ripped through the Cockshead seam, killing sixty-four miners and sixteen horses. Many of the victims are buried in Longton Cemetery. Then, on 21 March 1940, a further explosion killed eleven miners.

The colliery finally closed in 1963. The four pit wheels, carrying the names of coal seams and some neighbouring pits, are positioned in the levelled-off spoil heap at the Mossfield Colliery site.

The Mossfield disaster memorial at Longton Cemetery. (Author's collection)

> In memory of
> THE 64 MEN AND YOUTHS
> WHO LOST THEIR LIVES
> AT THE MOSSFIELD COLLIERY EXPLOSION
> ON OCT. 16TH 1889

The inscription on the memorial to those who died at Mossfield. (Author's collection)

> FRANCIS EMERY
> JOHN WILLIAMS
> WILLIAM LAWTON
> WILLIAM HULME
> NOAH BALL
> JOHN BALL
> JOSEPH BULL
> JOB BULL
> DAVID HUGHES
> JACOB BATH
>
> JOHN HALL
> GEORGE RATCLIFFE
> JOHN MOFFETT
> HENRY CALLCUTT
> HERBERT SELLARS
> ISAAC DERRICOTT
> THOMAS BROUGH
> EDWARD TOWNSEND
> EVAN PRICE
> THOMAS BRADSHAW
> BENJAMIN L. SMITH
> GEORGE SALT

The plaque naming the miners who are buried at Longton Cemetery. (Author's collection)

The four pit wheels from Mossfield as a memorial to the disaster of 16 October 1889. Plates on the wheels carry the names of local coal seams and pits. (Author's collection)

The tablet next to the Mossfield wheels: 'There's black ash beneath the green, Gob fires burning under Gas Hill, And beneath it all, Bones of miners buried still'. (Author's collection)

New Haden

The Draycott Collieries Company began work sinking New Haden Colliery, Cheadle, in February 1893. Known then as 'Klondyke', it operated in conjunction with the Draycott Cross and Delphouse workings and lay on the side of a sandstone ridge 1 mile south-west of Cheadle. To the side and below ran the 3-mile-long Cresswell–Cheadle branch line of the Cheadle Railway, which connected to the Stoke–Uttoxeter line on the North Staffordshire Railway. During the following thirteen years the mine experienced a succession of owners until it was bought by Staffordshire ironmasters Bansano Brothers Limited in December 1906. Then, in 1910, a new company, New Haden Collieries Limited, was formed to manage the mine, which had two main winding shafts: No. 7 Downcast was 12 feet 6 inches in diameter, while No. 8 Upcast was 8 feet 6 inches in diameter.

New Haden's fortunes began to decrease in November 1918 when a 400-foot section of railway tunnel roof fell onto the track, resulting in a month's closure of the line. Then, by 1933 the London, Midland & Scottish Railway had built a new line to Cheadle, which diverted around the tunnel, leaving the mine with a siding for access.

Work did begin, though, to upgrade the colliery in January 1927, with No. 8 shaft widened to 14 feet in diameter and deepened towards the Woodhead seam, where on 5 January 1927 it reached a final depth of 340 yards. Also, new headgear was installed together with a Markham of Chesterfield winding engine.

However, the mine was subject to water problems, and following efforts in 1942 to dig the 4-foot-deep coal, which in turn led to greater water flow, the Ministry of Fuel and Power closed the mine in 1943 as it was considered uneconomical in aiding the war effort. The Markham Winder was purchased by a pit in Hollinwood, Oldham, while the miners were transferred to Hem Heath and Florence. Today the site is an industrial estate.

Norton

The village of Norton-in-the-Moors developed around coal. From the Middle Ages, local proprietors and Lords of the Manor worked the top seams to supply iron foundries at Norton Green, Milton and Cobridge. A year after the opening of the Biddulph Valley railway line in 1859, Norton Colliery at Ford Green opened for the purpose of supplying coal to nearby Ford Green furnaces, which were later purchased by Mr Heath around 1863.

On the afternoon of Saturday 24 February 1912, three pit inspectors and repairers were caught in an explosion, which killed one man and injured the other two, when a fireball 800 yards long raced along a 1 in 3 incline toward the bottom of the shaft. Had it happened a few hours earlier, then 500 men would have been working in the pit. As it was, fifty horses died in their underground stalls from carbon monoxide poisoning. In 1928 Robert Heath and Low Moor collapsed, leaving Ford Green to become the Norton part of Norton & Biddulph Collieries Limited – a company formed with the intention to develop coal mining after iron and steel production had finished at Ford Green. Norton closed in 1977.

Oldfield

Oldfield Colliery, Fenton, dates from around 1826, when it was operated by William Hanbury Sparrow along with the Lane End Ironworks. There were two shafts with a depth of around 300 and 400 yards and an incline of 600 yards, to the south of which the workings extended for about 180 yards. On Friday 24 May 1855, nine men and one boy aged seventeen descended the shaft at 6 a.m. to repair the pit bottom engine. Then, between 9 and 10 a.m., an underground explosion so powerful that it lifted the cast-iron plates at the top of the shaft killed seven of the men. The cause was the naked flame carried by the boy igniting gas.

An earlier accident on 21 July 1843 saw the pit walls collapse inwards, trapping seven men and a horse. The men were able to escape through a narrow passage. Three days later, three men returned the same way with hay and water for the trapped horse, and again two days later. On 11 September the debris was cleared and the horse was found alive and rescued. It had been there some fifteen days.

By 1868 Oldfield passed to the Goddards, and by the early 1880s to the Balfour Group, being managed under the personal supervision of Jabez Balfour. Lane End Works Limited were running the pit by 1889 and closed part of it in 1891 before ownership passed to the Oldfield Colliery Company in 1896. This was short-lived as work stopped at the unprofitable mine on 14 September 1896, with it closing for good on Friday 18 September that year with a loss of 348 jobs.

Parkhall

Parkhall was initially known as the Western Coyney Colliery and had its beginnings around the 1850s, when it was owned by the Western Coyney & Cinderhill Colliery Company. By 1890 a rope-hauled tramway had been built from the pit to a landsale wharf on the site of Meir Hay Colliery, Longton. In the early twentieth century, the company went into liquidation and was subsequently bought by the Mossfield Colliery in 1908, with the mine being reopened. Later, a new washery was built in the 1920s, followed by pithead baths in the mid-1930s.

After nationalisation, Parkhall found itself part of the No. 1 North Staffordshire Area of the West Midlands Division of the National Coal Board; then, in the early 1960s, an underground connection joined the pit to Florence Colliery before Parkhall was finally closed in 1962. Interestingly, the rope-hauled tramway was still there when the mine closed. The landsale wharf – later known as Kendrick Street – was in operation until the 1990s.

Parkhouse

Parkhouse Colliery, Chesterton, was sunk in 1874 by Stanier & Company, who also owned mines at nearby Crackley. Originally the colliery comprised tandem mines for working ironstone, and under Stanier's system Parkhouse contained shafts Nos 6 and 7. No. 6, a drawing shaft, had a 9-foot diameter and was 328 yards deep, while No. 7, a drawing and pumping shaft, was 10 feet 6 inches in diameter with a depth of 328 yards. Around 1890 the mine was worked by J. H. Pearson, a Black Country ironmaster, until 1906, when Robert Heath & Sons Limited took over. From 1928 the pit was run by Parkhouse Collieries Limited before nationalisation in 1947. Coal winding stopped on 21 January 1968 and the pit closed in May of that year. The site is now an industrial park.

Parkhouse Colliery.

The site of Parkhouse Colliery. The track in the foreground is the route of the former railway. It is now an industrial estate. (Author's collection)

Racecourse

In the early nineteenth century, the Potteries Race Committee leased 47 acres of land from Wedgwood to develop a racecourse behind Etruria Hall. The first horse race was in 1824 and, supported by the Davenport and Copeland families, the enterprise proved to be a big success. By the 1830s, however, following the general election of June 1841, when both families withdrew their support, the racecourse project failed. The last race had taken place in 1840.

Following the demise of the racecourse, the 1st Earl Granville leased the mineral rights from the Duchy of Lancaster and sank a mine on the site to the west of Cobridge Road. In all there were three Racecourse Colliery pits, each almost 1,000 feet deep, which were owned along with The Grange, Rowhurst Nos 1 and 2, Boothern Pit and Tinkersclough. In the mid-1890s the sites employed more than 800 men and boys.

One tragic event happened at Racecourse Colliery on Thursday 20 August 1896, when miner George Higginson – who on the previous day had been injured by a cow! – complained of feeling unwell underground. Albert Lawton offered to accompany him to the surface but just below the top of the shaft both men fell out of the cage and were killed. It was assumed that Higginson had fainted and pulled Lawton with him.

The pits were known to be very wet and eventually flooded. They closed in 1937.

The Racecourse Pit.

Silverdale

Sir Robert Eggerton first discovered fossils in coal shale in 1763 and on 6 April 1792 the Silverdale Company was formed – the earliest recorded use of the name, predating the village of Silverdale – to excavate the Silverdale and Leycett areas, though by around 1815 the company had closed. As the landowner, Walter Sneyd of Keele Hall took over the pit until his death in 1929, when the Sneyd Estate Company was inherited by his son Ralph and run by agent Samuel Peake.

Later, in 1830, a coal mine supplied Ralph Sneyd's ironworks at Knutton Heath, which were situated around 300 yards north-west of the shaft. Samuel Peake died in 1848, leaving the company's affairs in disarray, and by that December a new company (the Silverdale Company) was founded, which was leased to Francis Stanier. Together, Sneyd and Stanier built the first standard gauge railway in North Staffordshire to the Pool Dam area of Newcastle. Known as the Silverdale & Newcastle-under-Lyme Railway, it was completed in 1850. Under Stanier and his son, Stanier-Broade (the company) prospered, having formed a partnership with Robert Heath.

Before Silverdale's main shaft was sunk, Ralph Sneyd died in July 1870. Then, by the 1880s the company was leased to Butterley Co. of Derbyshire, who from 1875 to 1881 worked Apedale, Park House (as oppose to Parkhouse Pit), Crackley, Knutton, Oak and The Grove before their lease expired in January 1901. A new company, the Silverdale Colliery Co. Ltd, closed the ironworks, forges and mills while working the remaining coal mines with limited success until 1918, when their property was sold to the Shelton Iron & Steel Company, who leased the mineral rights from Colonel Ralph Sneyd.

Keele Hall, home of the Sneyd family and owners of Silverdale Colliery.

The North Staffordshire Coalfield

Later, in the 1970s, the underground workings were connected via a 1 in 4 gradient surface drift mine, while a new coal field was discovered to the south by extending the drift a further 2,500 yards to a new pit bottom built under the M6 Keele services. The site, however, was still owned by the Sneyd Estate Trustees, who only sold it to the NCB *c.* 1980. The last NCB coal was drawn in December 1993 at the last deep mine in North Staffordshire. In 1994 Silverdale was leased by Coal Investment and produced coal until finally closing on 31 December 1998, thus ending seven centuries of mining in Staffordshire. The site is now a country park and housing estate.

A collier's horse and rider working in haulage at Silverdale.

NCB colliery gates at the former Silverdale Colliery. (Author's collection)

A monument to the miners of Silverdale Colliery, located in the village. (Author's collection)

Sneyd

Sneyd Colliery sat between Burslem and Smallthorne, where coal was mined in the eighteenth century, if not earlier. By 1851, C. & J. May held the mining lease. On 30 May 1874, a meeting took place at the Railway Hotel, Stoke, to form a limited liability company to manage Sneyd Colliery, which was known as J. Heath & Brothers. The 160-acre site also included fire-brick, tile quarry, marl and lime works. Initially there were three shafts, though by 1900 Sneyd Colliery owned the mining rights, sinking a new No. 4 shaft. In 1924 Nos 2 and 4 shafts were lifting coal while No. 1 was the upcast and manriding shaft. Meanwhile, No. 3 shaft was for access only, being later abandoned and filled in. Between the two world wars, Sneyd was one of the most modern mines in North Staffordshire, with coal wagons being taken to join the Potteries Loop Line at Burslem.

Among miners it was considered unlucky to dig coal on New Year's Day, and traditionally they didn't work. However, on 1 January 1942, to help the war effort, the Sneyd colliers arrived to work a normal shift. At 7.50 a.m. an explosion so powerful that it blew men off their feet occurred in the Banbury seam. In all, fifty-five miners were killed and two later died of their injuries. My grandfather, Albert Cooper, helped with the rescue.

Sneyd Colliery.

The Sneyd Colliery memorial in Burslem, remembering the disaster of 1 January 1942. (Author's collection)

The names of the miners who lost their lives at Sneyd Colliery. (Author's collection)

The plaque dedicated to the rescue teams at Sneyd Colliery. (Author's collection)

In the 1950s Sneyd was connected underground with Wolstanton. Coal winding later finished in July 1962, though for a while the shafts continued to be used for manriding. No. 4 shaft was to remain as an exit from the northern area of Wolstanton, however, until the latter's closure in 1985. Stoke-on-Trent College and Sneyd Industrial Estate now occupy the site.

Stafford

Stafford Colliery at Sideway, Great Fenton, began in 1873, initially to extract blackstone, ironstone and coal in the upper seams. Later, the Homer and Sutherland mines were sunk by C. J. Homer on behalf of the Duke of Sutherland in around 1875. The Sutherland shaft was 3,318 feet deep. Both mines had particularly ornate brick and stone engine houses complete with matching square chimneys. On Wednesday 8 April 1885, the Homer colliery suffered an explosion which killed four men and injured five.

The Duke of Sutherland, together with Messrs Pender, Charles Homer and John Bourne, owned three pits by 1884 and a further two by 1891. In 1924 Stafford became the coal processing plant for the recently opened Hem Heath Colliery to the south, with three of the five Stafford mines still operating in 1957. Later, in the early 1960s, the decision was reversed, with Stafford's coal preparation facilities being concentrated at Hem Heath. Stafford eventually closed in 1969.

Stafford Colliery.

Stoke City's stadium on the site of Stafford Colliery. (Author's collection)

Talk O' Th' Hill

Talk O' Th' Hill – a colloquial name used locally for 'Talke on the Hill' – was a colliery which lay on the west side of the North Staffordshire Coalfield. The largest share of the Talke coalfield was owned by the Sneyd family of Keele Hall, though in the 1790s mining rights were given to John Gilbert of Kidsgrove, while later, in 1809, Thomas Kinnersley of Clough Hall obtained further grants.

Talk O' Th' Hill had three shafts: No. 1 was 11 feet 6 inches in diameter and was an upcast shaft fitted for winding; 65 yards away, No. 2 was the main coal-drawing shaft, and was 999 feet deep with a diameter of 11 feet 6 inches; No. 3 shaft lay some 600 yards to the north-west of the former two and was 765 feet deep.

The colliery suffered several explosions throughout its history. An account from 1866 reads:

> Thirteen men and boys were rescued alive, two of whom died later. By Friday morning fifty-eight bodies had been recovered and removed to 'The Swan Inn'. Two rooms had been provided there for the reception of bodies and identification by relatives. There were terrible scenes as people searched for their loved ones, many of whom were dreadfully mutilated. The cause of the explosion was never established, but several

The Swan Inn, Talke, where the deceased miners' bodies were taken following the 1873 explosion at Talke Colliery. (Author's collection)

homemade keys to the safety lamps were found in the pockets of those who died. Evidence was also found of miners smoking underground. A collection ordered by Queen Victoria raised over £16,000 for the widows and children.

Then, in 1873, twelve months after the North Staffordshire Coal & Iron Works opened, ninety-one workers were killed, the youngest of whom was ten years old. Furthermore, at 7.30 p.m. on 27 May 1901 (Whit Monday), an explosion resulting from a gob fire in the 8-foot Banbury seam killed four men and twenty-seven horses and ponies. Had it not been a holiday, even more lives would have been lost.

Confusingly, the site is now home to the Jamage Industrial Estate.

Ubberley

Ubberley Colliery was located between Bucknall and Hanley. Encompassing a number of mines, the principal ones were the Sampson Pits. As early as 1834 it was listed in White's Directory as belonging to John Ridgeway Esq. of Cauldon Place, and in 1836 Church Rates were paid to Bucknall Church.

On the morning of 25 August 1851, thirteen men and boys descended the engine shaft at the main Red Pit in two wagons between 6 and 7 a.m. The first six, who all had safety lamps, went to the north side, closing the doors after them, while the following seven entered the south mine. Only one, Ralph Hancock, had a safety lamp, while those following carried lit candles. Within half an hour, and barely 300 yards into the workings, a huge explosion killed the men instantly. Those on the north side survived. Poor ventilation in warm weather had led to a build-up of gas, which was lit by the naked flames. My three times great-grandfather, Edward Forrester, a turnsman, was one of the men killed. The story goes that he was late for work that morning and ran to make the wagon just as it was descending. He left a widow and seven children.

Ubberley may have been bought by C. J. H. Homer before 1864, when the Chatterley Iron Company was formed. Sometime before 1875, when he left the company, Homer installed a vertical winding engine. By 1881 the colliery was owned by Chatterley Whitfield Collieries Limited, though probably because of the earlier refurbishment it was known as New Ubberley. The colliery closed sometime around 1904.

Ubberley Pit.

Edward Forrester, my three times great-grandfather, who died in the Ubberley Pit disaster. (Author's collection)

Victoria

Victoria Colliery, Biddulph, began in 1850 and was known as 'The Bull' or 'Black Bull' after the nearby village of that name.

John Bateman of nearby Knypersley Hall leased the mineral rights under the estate to Robert Heath in 1857, where three shallow shafts were already in place. Heath went on to develop the Biddulph Valley Colliery based on the existing shafts, and for the majority of its productive life the pit was to have three shafts. No. 2 shaft was deepened and known as the Magpie shaft, while two new shafts shared one winding engine and were later known as the Deep Pit and Bye Pit. Following further excavation, the two went on to become the Havelock shaft.

By 1887 Robert Heath had acquired the Brown Lees Pit, where between 1883 and 1885 a shaft was deepened to enable an underground connection with the Biddulph Valley Pit, while the Salisbury pumping shaft was sunk at Black Bull to a depth of 240 yards. The old shafts at Brown Lees also became pumping shafts. Later, in 1898, work began to widen and deepen the Magpie shaft. Finally completed in 1900, the mine was renamed Victoria.

Difficult geological conditions meant that the colliery closed in 1982 and the pumping shafts at Brown Lees were filled in. Afterwards, the rope pulley wheels – measuring 6 feet in diameter – were donated to Biddulph Council and were subsequently erected by the town hall. However, opencast mining continued on the site through the late 1980s and early 1990s. On the Brown Lees site, many remains of the old Black Bull & Biddulph Valley Ironworks (also owned by Robert Heath) along with the pit shafts were later discovered. Of particular interest was the finding of a very early steam engine that initially came from Turnhurst Pit and was bought for Black Bull in 1886.

Knypersley Hall, the home of John Bateman, owner of Victoria Colliery.

Left: My father, Philip Cooper, at Victoria Colliery *c.* 1978. (Author's collection)

Below: The balance rope pulley wheel from Victoria Colliery outside Biddulph Town Hall. (Author's collection)

Above: The plaque on the monument to Victoria Colliery. (Author's collection)

Right: The Brown Lees village crest, depicting a loco on the Biddulph Valley Line and Victoria Pit's winding gear.

Wolstanton

Wolstanton Colliery was sunk by pottery manufacturers in 1920 on land owned by the Duchy of Lancaster, which only came into National Coal Board ownership in the 1980s. Its two shafts were originally sunk to work the ironstone that sat on top of a viable coal seam; later, in 1927, No. 1 shaft was deepened to 635 yards to access the coal. In the 1950s No. 2 shaft was deepened to 1,140 yards and in May 1957 a third shaft was sunk, making Wolstanton the deepest pit in Europe. In October that year equipment was installed to take methane gas from the coal face to Etruria Gasworks. Following treatment, it was then piped into homes and factories, such as Michelin, H. & R. Johnson and Downing's.

Following the abandonment of plans for a colliery at nearby Bradwell, Wolstanton, due to its access to the railway line, was joined underground to Sneyd and Hanley Deep in 1962, Norton in 1968 and Chatterley Whitfield in 1974. Later, during the miners' strike of 1984, my father was allowed to cross the picket line in order to feed the feral cats living there.

The colliery was officially closed on 14 March 1986, although it actually closed earlier, on 18 October 1985. Work began on 17 March 1986 to fill in the shafts. Today the site is home to a retail park.

Wolstanton Colliery. (Author's collection)

The pithead at Wolstanton Colliery in the 1980s. (Author's collection)

Inside the offices at Wolstanton Colliery. (Author's collection)

A shunting engine at Wolstanton Colliery. (Author's collection)

The colliery bridge over the A500, which carried coal by conveyer belt over the road prior to demolition in October 1987. Under the bridge to the left of the northbound carriageway was a sign that read: 'Keep your eyes on the road'. (Author's collection)

Wolstanton No. 3. (Author's collection)

A 1984 plaque celebrating Wolstanton Wakes, depicting St Margaret's Church, the colliery and Haley's Comet. (Author's collection)

The South Staffordshire Coalfield

Baggeridge

In February 1899 work began on sinking Baggeridge Colliery's first shaft and in July 1902 it reached a 24-foot-thick coal seam with a total depth of 2,019 and a diameter of 17 feet. A second shaft into the same seam, which was 1,929 feet deep, followed in 1910, though this was plagued with water problems, and a 40-yard brick lining (called 'coffering') was installed to keep the water at bay. In total the cost was £25,000, and coal production began in 1910.

Baggeridge was owned by the Earl of Dudley from nearby Himley Hall, and although geographically it lay just outside the South Staffordshire Coalfield, it is still within the county despite the 1974 boundary changes, and was the last major attempt to mine the 'Thick Coal' of the Black Country. The pit was served by the Earl's private Pensnett Railway, which was built by the Great Western Railway in 1907, with another line connecting to Ashwood Basin on the canal. Interestingly, there was also a third shaft sunk, by which the South Staffordshire Water Company pumped water to nearby Sedgley Reservoir.

Baggeridge closed on Sunday 2 March 1968, and was the last mine in operation in that area at the time. The site is now Baggeridge Country Park, which is open to the public.

Brereton

Brereton Colliery was the most productive and sole survivor of several shafts sunk between 1817 and 1820 by the Marquis of Anglesey. Originally built as an additional shaft of Hayes Colliery, it opened in the 1820s and between 1828 and 1847 was leased as the Hayes Collieries by Joseph Palmer and subsequently his widow, Sarah. However, due to management failure the Marquis regained control in 1847. In 1854, following temporary closure, the lease transferred to Earl Talbot, along with mineral rights over 2,000 acres.

Above: Sculptures at the former pithead in Baggeridge Country Park standing on the concrete foundations of the winding gear, designed by Steve Field and made by D. R. Harvey of Dudley. (Author's collection)

Right: The 1st Earl of Dudley.

Himley Hall, home of the Earls of Dudley and later NCB West Midlands' divisional headquarters. (Author's collection)

The crest of the National Coal Board. (Author's collection)

Brereton Colliery.

By now the mine had two 8-foot-diameter shafts: the Brick Kiln shaft was 255 feet deep while the Engine shaft had a depth of 105 feet. By 1883 Earl Talbot had redeveloped the mine with a new 467-foot shaft, 15 feet in diameter, while also deepening one of the old shafts to 466 feet with a diameter of 14 feet, which became known as the Ingestre shaft. In 1906 there was further expansion under the name of Brereton Collieries Limited before the mine's new owner, James Cadman, took control in 1920. Development continued in 1925–6 when a 60-foot-high concrete pithead frame was cast on site, encasing the unsafe wooden one.

By 1960 Brereton had become uneconomic and the last coal train left the mine in July of that year.

Cannock & Leacroft

On Saturday 5 April 1873, the Cannock & Leacroft Colliery Company began surveying and marking out the land leased from J. S. Gibbons for their new mine, known as 'Leacroft', which lay close to the London & North Western Railway's Norton line and in the vicinity of the East Cannock coal wharf's extension from the Wyrley & Essington Canal. In 1874 work began on the two shafts, which were around 101 yards apart. No. 1 had a depth of 407 yards with a diameter of 13 feet 4 inches while No. 2 was 389 yards deep and 12 feet 6 inches in diameter. They were completed in 1877.

During the 1920s a distillation plant was opened by the Patent Economic Fuel Company to process liquid fuel from Leacroft's coal, but it proved a failure and the colliery took the plant over as payment for coal before closing it in 1929. Later, in 1933,

Leacroft Colliery.

it reopened under Mr W. B. Midford, who succeeded in developing petrol from coal and a car was subsequently fuelled by the first petrol made from coal in Great Britain. The plant, however, had closed by the 1940s. Cannock & Leacroft Colliery had little modernisation prior to being nationalised and afterwards continued to be thought of as old fashioned. Indeed, there were no pithead baths and in the latter years miners were taken by bus to Mid Cannock Colliery to use their facilities. The colliery closed in 1954 following its underground connection with Mid Cannock Colliery.

Cannock Chase No. 1 Colliery
Begun in 1848 by the Marquis of Anglesey, the mine was originally known as Hammerwich Colliery until 1859, when John Robinson McClean leased the pit from the Marquis and formed the Cannock Chase Colliery Company. Hammerwich was their first mine, hence the later name. It lay close to Norton Pool (now the Chasewater retaining dam) with the pithead below the water level! Hammerwich had three shafts and was opened in 1849 No. 1 was 12 feet in diameter and No. 2 was 10 feet in diameter, while No. 3, a ventilation shaft was 6 feet in diameter. The last coal was drawn at the end of January 1857 as an inrush of sand and gravel forced the mine to close. Interestingly, after closure the colliery engine – essentially a beam engine – was transferred to No. 7 Colliery.

Chasewater Dam, the site of Cannock No. 1 Colliery, originally known as Hammerwich. (Author's collection)

Cannock Chase No. 2 Colliery

On 3 May 1852 the Marquis of Anglesey opened the originally named Uxbridge Colliery (dedicated to himself, as he also held the title of Earl of Uxbridge) though it was known locally as 'The Fly' due to the speed of its winding. The pit had two shafts, which were each 9 feet in diameter, while No. 1 Upcast was 360 feet deep and No. 2 Downcast was 361 feet deep. In 1854 Uxbridge was taken over by the Cannock Chase Colliery Company and became Cannock Chase No. 2 Colliery. Meanwhile, the shafts were deepened to 530 metres and the site was developed to include offices, workshops and locomotive sheds for the company.

Following the closure of No. 1 Colliery, Uxbridge extracted its reserves. In 1923 a 945-yard drift gave access to the workings from an opening near Anglesey Wharf and the nearby railway, making the shafts redundant. As a result, permission was granted to mine the reserves in the shaft pillar while the remaining coal was worked via No. 3 Colliery. Uxbridge closed in 1940 and the site is now home to Burntwood Rugby Club.

The Marquis of Anglesey, owner of Cannock Nos 1 and 2 Collieries. (Black Country Museums)

Cannock Chase No. 3 Colliery

No. 3 Colliery was known locally as 'The Plant', possibly due to the large quantity of new machinery installed there. Work began on the mine's two shafts in 1859 and two years later, at a cost of £5,175, the Downcast had a diameter of 13 feet and a depth of 489 feet while the Upcast shaft was 9 feet in diameter and 495 feet deep. The first coal surfaced in December 1861. The Plant replaced No. 2 Colliery in 1923 as the operation's centre, with new screening plants and a washery. In 1924 a large steel building called Wembley was erected to house the colliery workshops, costing £3,889. Prior to the Second World War, the pitch pine headframes were replaced with steel.

After nationalisation the workshops became the Area Central Workshops before going on to serve all regions from 1957 until their closure in 1990. The Plant closed in 1959 due to a lack of reserves and today the site is an industrial estate adjacent to the Burntwood Ring Road.

Cannock Chase No. 4 Colliery

Few records survive of No. 4 Colliery in Chasetown, though confusingly it was the Cannock Chase Company's third mine. It was possibly sunk by the Marquis of Anglesey after 1852; however, as the Cannock Chase Company was founded in 1854 and the first coal surfaced on 22 October 1856, its true origins are unclear. The mine had three shafts, one of which is thought to be for the ventilation of No. 2 Colliery's workings. The last coal to be drawn was in January 1877.

Cannock Chase No. 5 Colliery

No. 5 Colliery lay 1,000 yards from No. 3, close to Cannock Road. It opened in 1862 at a cost of £4,375 to develop the underground workings and the first coal was drawn in early September that year. The mine had three shafts: two were 9 feet in diameter with a depth of 366 feet, while the third was a 'bull pit', being used to drain water that in turn created steam power for the machinery. Due to the pit's close proximity to the Eastern Boundary Fault (excess water prohibited working through the fault) it was a one-sided mine, which meant that the coal lying between Nos 5 and 3 Collieries was shared by both pits. In 1907 the company built its power station at No. 5, and following the closure of the mine in 1920 it continued as part of the National Grid, bringing the first electricity supply to the neighbourhood in 1922. The power station remained open until the early 1940s, while some of the colliery's buildings survived into the 1960s.

Cannock Chase No. 6 Colliery

No. 6 Colliery was known as Cannel Mount and opened in 1866 between Heath Hayes and Wimblebury. It had two shafts to access the high-quality cannel coal approximately 30 yards below the surface, from which it was planned to distil oil or petrol. However, in 1867 independent analysts discovered that the coal produced 11,000 cubic metres of gas per ton, leading to the company building a gas plant in Chasetown. Despite early optimism, the mine proved unsustainable, and the colliery closed in 1874. The gas plant continued to supply the company's collieries and the surrounding area before being taken over by a gas company in the 1930s.

Cannock Chase No. 7 Colliery

The company's No. 7 Colliery was located in Prospect Village. Work began on the shafts in 1868, initially being completed by 10 December 1874 before the shafts were deepened further in 1894 and then again in 1907, ending up at 852 feet deep with a diameter of 9 feet. Unfortunately, the mine suffered major water problems, and as a result may not have achieved full production until the early twentieth century; even then the coal reserves were limited due to the nearby Eastern Boundary Fault and the neighbouring Cannock & Rugeley Colliery Company's land.

Coal was last drawn from No. 7 in 1923. Though miners and materials still descended the shafts, coal was surfaced via the drift at No. 2 Colliery. Production finished in 1936, but the shafts remained open for several years to ventilate the other mines and to pump water from the Eastern Boundary Fault.

Cannock Chase No. 8 Colliery

No. 8 Colliery, began in 1868, lay next to No. 6 and close to Cannel Mount. Both shafts were 12 feet 6 inches in diameter while the Downcast was already some 30 yards deep before Matthew Boothby & Company won the contract to sink the shaft a further 295 yards at a cost of around £5.71 per yard. Similarly, the Upcast was 330 yards deep at a cost or about £5.40 per yard. Also, a 130-yard-deep bull pit positioned between the other two shafts cost £2.75 per yard. In total, the underground workings cost £3,843 3s 7d (£3,843.18).

The mine was modernised in 1927–8 when a new headframe was built, together with a new winding engine house sited south of the Upcast shaft. Meanwhile, the old haulage engine was redeployed to drive the approximately 2,200-yard endless rope haulage tub tramway to No. 3's plant, taking the coal there for screening. Later, in the early 1950s, an underground link was made with Cannock No. 9; subsequently, the coal from both mines was wound at No. 8 Colliery. Cannock No. 8 was the company's only pit to have purpose-built pithead baths, with the miners paying a weekly contribution to use them. When it closed in 1962, the colliery was the last survivor of the company's original pits.

Cannock Chase Nos 9 and 10 Collieries

Cannock Chase No. 9, known as 'Old Hednesford', and No. 10 are linked as together they formed Hednesford Colliery. Their early history is somewhat complicated as both passed through the hands of several owners.

In the nineteenth century Messrs Webb & Poxon sank a shaft some 48 feet deep before the site was purchased by Mr Francis Piggott in the late 1850s, who excavated the workings further to reach the shallow seams. Later, on 1 September 1860, a contract was made between Francis Piggott and the Company of Proprietors of the Birmingham Canal Navigation, which duly led to the formation of the Hednesford Colliery Company in 1864, with Francis Piggott as Managing Director and a capital sum of £60,000. However, by 1867 Mr Piggott had relinquished the lease to Richard Byrd Levett due to poor output, who in turn leased the site to William Tredwell in 1869 for a period of forty-two years. Later, in 1870, John Robinson McClean agreed to buy the mine together with all the plant for £110,000 – a considerable sum of money, which was paid in instalments. Hednesford Colliery then became Cannock Chase Nos 9 and 10 Collieries, but this led to disagreements among the Board members and in 1877 Chairman Frank McClean resigned. By the early twentieth century No. 9 had become the larger of the two mines and No. 10 closed around 1910, its coal reserves being accessed from No. 9 Colliery. Moreover, it was not until 1916 that the Cannock Chase Colliery Company owned the mine in its entirety.

In 1911, Old Hednesford, or No. 9 Colliery, became the site of one of the worst mining disasters in Cannock Chase when around noon on Thursday 14 December five men died from suffocation. Some 160 men were working underground at the time and soon after their break, or 'snap time', a fire broke out in the lamphouse, located around 20 yards from the shaft bottom. There, it was common practice to store oil to refill the lamps, or 'shukeys', and as a consequence some of the oil would drip onto the ground. In addition, boys would trim their lamps and discard the oil-soaked wicks on the floor. It is thought that around 11.30 a.m. a boy did not extinguish his lamp before filling it, causing the spilled oil to ignite and the fire to spread rapidly. All but five who could not be reached managed to escape, and despite the heavy rain a large crowd assembled at the pithead, waiting for news. Later, Henry Merritt of Heath Hayes and the widow of Thomas Stokes received the Edward Medal, First Class, for bravery during the disaster.

Coal winding finished at No. 9 in 1927, but due to the changing market the remaining coal in the upper seams was later deemed suitable for gas and electricity generation and the mine was reopened in 1935. The shafts were reversed, presumably with a new headframe and engine installed, while the original wooden Upcast frame

Hednesford's Davey lamp memorial. The brickwork carries individual miners' names. It was unveiled on 30 July 2000. (Author's collection)

Staffordshire Coal Mines

The nearby garden at Hednesford, where local colliery names are engraved on paving stones and miners are remembered on decorative bricks. (Author's collection)

Some of the miners' names at Hednesford. (Author's collection)

The Cannock Wood stone at Hednesford. (Author's collection)

The West Cannock No. 5 stone at Hednesford. (Author's collection)

and winding engine were removed to No. 7 Colliery and continued in operation there until nationalisation. In the 1950s coal produced at Old Hednesford was transported to No. 8 via an underground endless rope haulage system, being brought to the surface there. The colliery finally closed in 1962.

Cannock Old Coppice

Old Coppice, later known as Hawkins Colliery, lay in Cheslyn Hay. William Gilpin began mining the area in 1817 before Edward Sayer, a local coal-master, sank two shafts on the site around 1848. In 1869 Joseph Hawkins took over the colliery's lease and a report on Friday 7 April 1876 announced that the Cannock & Old Coppice Colliery Company had been registered with a capital of £100,000 in £100 shares 'for acquiring and working pits in the parish of Cheslyn Hay'. Joseph Hawkins owned the mine until his death in 1907, aged ninety-four, when ownership passed to his widow and sons. Between 1876 and 1878 No. 1 shaft was deepened to 280 yards with a diameter of 12 feet while No. 2 (Upcast) was sunk to a depth of 1,185 feet, with the diameter being enlarged from 9 feet to 15 feet. The mine had a narrow gauge tramway, built around 1880, linking it to a wharf on the Staffordshire & Worcester Canal, followed by the construction of a purpose-built canal basin in 1883 at Bridgetown. Later, around 1900, a mineral line connected the pit to the London & North Western Railway to the south of Churchbridge.

On Monday 3 September 1900, between 6 and 7 a.m., some seventy miners had already been lowered in the cage to start the dayshift. When the cage descended again with eight men on board, the engineman, William Hill, lost control and the cage hurtled to the bottom of the shaft. Two men were thrown from the cage (there were no gates on the cage), which rebounded and threw out the other six. All fell to the bottom and were crushed by the falling cage. Hill claimed that the winding engine lever had stuck but was later charged with manslaughter, before being found not guilty and acquitted.

After nationalisation Old Coppice was officially known as Hawkins Colliery after the previous owners. It was closed by the National Coal Board in April 1960 as being uneconomical.

Cannock Wood Colliery

Cannock Wood Colliery, also known as the Wood Pit, Rawnsley Pit or Cannock Wood, was the highest pithead on Cannock Chase at 205 metres above sea level. Begun in 1864 by the Cannock & Rugley Colliery Company, it lay 1½ miles north of Hednesford. By November work on the two shafts, each 12 feet in diameter and 120 yards apart, was progressing well. However, water had become a problem, and in order to protect the workings a 6-foot-diameter well shaft was sunk as a drain between the two shafts. The mine was producing coal by February 1867. By 1872/3 it had grown larger and another shaft, known as the Vent shaft, 16 feet in diameter, was sunk to expel the return air and carry men and materials underground.

The Cannock & Rugeley Colliery Company was always willing to embrace new technology, so when Colliery Manager Mr Williamson invented a safety oil lamp, which he patented in 1877, all of the company's miners had a Williamson lamp by the 1890s.

At nationalisation Cannock Wood was the most productive mine on Cannock Chase and so between 1957 and 1962 the National Coal Board invested heavily in modernisation. Moreover, throughout the work production targets were still maintained. Seven years later, in 1969, the last of the pit ponies working in the Cannock Chase coalfield were retired from the Wood. The mine finally closed on 8 June 1973. The site is now an industrial estate.

Conduit No. 3 Colliery

Conduit No. 3, known as 'Jerome's Pit', was one of the mines owned by the Reverend Jerome Clapp Jerome, the father of novelist Jerome K. Jerome, in the vicinity of Watling Street or the A5. Conduit No. 1 had three shafts on Wyrley Common from 1864 to 1897, while No. 2, also called the 'Wilkin Pit', had a shaft close to Wilkin Road known as the Corner Pit and two more shafts, known as the Engine Pits, which lay some 500 yards to the north-west.

Work began on two shafts in 1858, almost in the centre of Norton Canes village. However, due to adverse conditions of shifting sand and flooding, it was some years until production could begin. This proved to be a bad investment for the Jerome family and left them virtually penniless, resulting in the mines, plant and machinery being sold to the original landowners, William Hanbury and his son-in-law R. W. Hanbury Masfen, in 1860. In turn, they were purchased for a few hundred pounds by James and Charles Holdcroft, who finished developing the pit, with the shafts becoming Nos 9 and 10 of the Conduit Colliery Company (founded in 1864). During the late 1880s two more shafts were sunk, No. 11 and the Upcast shaft, both close to the original ones.

The Conduit Colliery at Norton Cains.

In the early 1900s the company acquired Norton Green Pit. Sunk by William Harrison in 1874 and initially closed in the late 1890s, it was reopened in 1903 by the Conduit Colliery Company as their No. 4 Colliery.

On 1 January 1931, the Littleton Collieries Limited took over Conduit Nos 3 and 4 Collieries and subsequently closed No. 4 in July 1933, with its reserves being accessed via an underground link with Jerome's Pit. Later, following nationalisation in 1947, the National Coal Board decided not to update Jerome's as reserves were running low. The last coal was lifted on 29 July 1949.

Coppice Colliery

Coppice Colliery in Heath Hayes began in 1893 when the Right Honourable William Hanbury, Member of Parliament and former Minister of Agriculture, from Ilam Hall, Derbyshire, sank two shafts on his own land. His wife, Lady Bowering-Hanbury, cut the first turf and later went on to be the mine's coal-master, being one of the few women to do so. It would take two years to complete the shafts, with the first coal being produced in 1895.

Following her husband's death in 1903, the colliery was run by his trustee, Victor Henry Bowering-Hanbury. However, Lady Bowering-Hanbury continued to pay frequent visits to her pit and was a popular figure among the miners, showing concern for their welfare. In her honour, the colliery became known locally as 'The Fair Lady'. When she died in 1931 a new company, the Coppice Colliery Company, owned by Lady

The Coppice Colliery at Heath Hayes.

Ilam Hall, the former home of the Right Honourable William Hanbury, owner of Coppice Colliery.

Bowering-Hanbury's six nieces, took over. It was probably unique for an all-female family to be owning a coal mine. The colliery workers knew them as 'the nieces'.

After nationalisation, the NCB planned its first improvement – to replace No. 2 Upcast shaft's rotting wooden headframe with the one from Conduit No. 4 Colliery's Upcast shaft during the summer of 1948. Then, in 1953, prefabricated pithead baths, formerly of Yew Tree Drift site, were installed to the rear of the lamproom. By the early 1960s reserves were running low and Coppice finally closed on 25 April 1964.

East Cannock Colliery

The East Cannock Colliery lay close to the Globe Inn and next to the East Cannock Road in Hednesford. Founded on 13 December 1870, the East Cannock Colliery Company made plans to sink two shafts at a cost of £120,000. However, work did not begin until 1875, twelve months after the Vice-Chairman's daughter cut the first turf on 28 March 1874. Coal production eventually started in 1876, but not before No. 1 shaft was named 'Amy' after Amy Stokes, the Chairman's daughter.

In 1877, as the company's finances were improving, a methane explosion around 1.30 p.m. on 31 August 1877 resulted in the death of four boys, while another boy and workman suffered serious injuries. Four horses also died. The inquest recorded that the explosion in No. 1 level caused the victims to be burnt by ignited gas, ruling that it was accidental death.

Staffordshire Coal Mines

East Cannock Colliery.

East Cannock Colliery *c.* 1880.

By 1880 the mine's finances came to the fore again as the business failed. The mine was then bought by Henry Davis Pochin, a Welsh businessman, for £20,000 and the Second East Cannock Colliery Company was founded on 2 November 1880. Following nationalisation in 1947, East Cannock was never part of the National Coal Board's future plans, and it closed in 1957 due to a lack of reserves.

Fair Oak Colliery

The Fair Oak Colliery Company Limited began with a capital sum of £200,000 on 1 September 1871 before acquiring a sixty-year mineral rights lease on Cannock Chase from the Marquis of Anglesey on 10 October that year. The company took its name from a huge oak tree growing in the centre of their land. On 1 January 1872 the first turf of Nos 1 and 2 shafts was dug by Company Directors Major William Arden of Longcroft House, Yoxall, and Mr W. H. Moore of Kidderminster. However, after nearly four years of hard work the excavations were abandoned on 1 December 1875 due to persistent flooding and a new site 1½ miles to the west and close to the Fair Oak Tree was selected. No. 2 Plant, with shafts Nos 3 and 4, began on 24 October 1876. Eventually, in autumn 1877, with the shafts at a depth of 137 yards, the mine at last started to produce coal and the company had some income.

Later, in early 1884, the company was doing well and had plans for a No. 3 Plant around 2 miles distant on Brocton Field. Unbelievably though for the Fair Oaks Company, on Tuesday 25 June a single creditor, George Craddock of Wakefield, presented a winding-up petition over a comparatively small amount of money. The Fair Oaks Colliery closed on 9 July 1884 when the company went into liquidation. There were some attempts to reopen the pit but these proved unsuccessful and the shafts were covered. It was not until the 1950s that the National Coal Board filled in the shafts and capped them with concrete. The site is now managed by the Forestry Commission.

Great Wyrley

Great Wyrley Colliery sat next to the London & North Western railway, now the Walsall to Rugeley line, just north of Landywood station. It was known locally as 'The Plant' to avoid confusion with Harrison's Brownhills No. 3 Colliery. In 1871 the Great Wyrley Colliery Company was formed and work began on the New Plant Colliery on 9 May 1876. The two shafts – both 204 yards deep and 45 yards apart – were completed on 9 January 1878, while in August 1880 the railway and surface buildings were ready for production to begin. The company based their centre of operations in the new offices. Unusually, there were no underground stables, so the ponies were bought to the surface each day.

Meanwhile, water proved to be a problem, with the mine flooding in 1905 and again on 29 July 1910, when the men and fifty horses were evacuated. One pony refused to pass through the water and was left on a higher roadway in complete darkness with food and water. Five days later he was found fit and well.

The mine proved to be successful and earned the name 'The Golden Mine'. As a result of the depression, however, the company went into liquidation on 21 April 1925, with the mine closing. However, the Great Wyrley Colliery was reopened by Charles Screen of Oldbury's company, Nook & Wyrley Collieries Limited, around 1926. Following nationalisation, the NCB considered the pit to be uneconomic and it closed again in August 1949.

Mid Cannock Colliery

The Mid Cannock Colliery Company (Limited) was formed in 1872 to sink a mine – known as 'The Mid' – around 1 kilometre south of Cannock town centre on land leased from the Marquis of Anglesey, who had the freehold, and Lord Hathern, who was the copyholder. Sinking of the two shafts, which were about 99 yards apart, began in 1876 and production started as soon as the shafts were completed. Hard times saw the company fall into voluntary liquidation in August 1880 before a new company – the Mid Cannock Colliery Company Limited – took over on 11 July 1881. However, due to the economic depression of the period, along with water problems, the mine closed and lay abandoned for twenty-nine years, during which time the forward-thinking William Harrison rented the land. When conditions improved William Harrison Limited purchased the site for £45,000 in 1913 – less than half of the original cost to sink the mine, even though the water level in both shafts was around 12 yards from the surface. A year later coal was being drawn again and demand was high due to the need to support the war effort during the First World War.

On 30 November 1915 an explosion caused by the fireman detonating an explosive charge killed five miners, two of whom died at the scene and the others later in hospital due to severe burns. A verdict of accidental death was recorded.

In September 1954 an underground connection with Cannock & Leacroft Colliery was finished and later, in April 1963, a similar roadway connection with Wyrley No. 3 Colliery was completed. Henceforth all the coal was hauled at Mid Cannock. Mid Cannock and its amalgamated pits closed in December 1967 due to a lack of economic reserves.

The Nook

The Nook Colliery lay on the southern edge of Cheslyn Hay next to a canal basin off the Wyrley & Essington Canal. The shafts were 45 yards apart and 220 yards deep. The pit began around 1863 and early documents record it as belonging to a Mr Walton. However, by December 1873 the Wyrley Cannock Colliery Company was leasing the mineral rights from the Duke of Sutherland. By 1924 the mine was becoming unprofitable and it closed on 21 April 1925, when the company went into liquidation.

Following the 1926 General Strike, The Nook Colliery was sold and reopened in October that year under the new ownership of Charles Screen of Screen Brothers Limited through their new company, The Nook & Wyrley Collieries Limited. Later, in 1939, economic reserves were running out and substantial difficulties with water led to the end of mining, although some work still continued above ground. The colliery finally closed for good on 30 June 1941.

Hilton Main

Hilton Main Colliery, known as 'The Main', began as an offshoot of Holly Bank Mine, Essington, when work on a new shaft began in 1919 by the Holly Bank Colliery Company. The shaft was completed in 1923 and officially opened in September 1924. It was 620 yards deep with a diameter of 18 feet while the steel headgear was 65 feet high. Initially it depended on Holly Bank for ventilation until excavation of a new shaft at Hilton Main began on 17 April 1934, a few yards from No. 1 shaft, and it plunged to a final depth of 360 yards under the new Hilton Main Colliery Company. On completion in 1936 No. 1 shaft became solely responsible for coal drawing and with a new engine house it was the only Midlands pit to have two electric winders. On 7 November that year Hilton Main took over the ventilation control from Holy Bank, making the mine independent. Moreover, on 17 June 1939 pithead baths were opened and the miners were charged 5d a week to use them. Hilton Main closed in 1969.

Holly Bank

Holly Bank was the successor of Essington Wood Colliery, which began in 1840 and was taken over by the newly formed Holly Bank Colliery Company in 1891. Coal from Essington Wood was transported to the Wyrley & Essington Canal via a tramway before a railway was built in 1875 to connect with the Walsall to Cannock line of the London & North Western Railway. Later, in 1906, another railway connected Holly

Railway engine *Holly Bank No. 3* from Holly Bank Colliery. (Author's collection)

Bank to the new canal wharf at Short Heath, bypassing the canal lock. In 1920 it was the first pit on Cannock Chase to sink a new shaft into a haulage road to access coal reserves, which due to a geological fault were too difficult to reach from the existing shaft. This became the site for Hilton Main.

Holly Bank first closed in the 1920s due to the recession and production transferred to Hilton Main as they were both owned by the Holly Bank Coal Company – though it was soon to be taken over on 1 July 1932 by Hartley Main Collieries Limited of Cramlington, Northumberland. In 1935 a new company was formed and, due to better economic conditions, Holly Bank Colliery reopened in 1938, later becoming a service shaft for Hilton Main and then a pumping shaft in 1952. However, the mine had problems with water and became uneconomical, leading to the National Coal Board announcing the closure of Holly Bank on 26 December 1952.

Lea Hall

Work began at Lea Hall Colliery, Rugeley, in 1954 when the two shafts were excavated close to the railway and the River Trent. At the time many neighbouring pits were ageing and uneconomic, and Lea Hall became the first mine to be opened by the National

The 9-foot statues that make up the memorial to the 115 miners killed at Lea Hall and Brereton Collieries on Globe Island, Rugeley, designed by Amy De Comyn and unveiled on 13 September 2015.

Coal Board at a cost of £14 million. The shafts were 396 metres deep, traveling through alluvium ground, which had been frozen during construction. In July 1960, the first coal was produced.

At the same time Rugeley A Power Station was built as a joint venture and coal was transported via a conveyor belt directly to the power station. Seven coal faces were in operation and the mine was the largest of its type in Europe, as well as being one of the world's most modern pits.

Lea Hall closed on 24 January 1991 and the site is now the 100-acre Towers Business Park.

This statue shows a miner prior to the 1930s, without safety gear. (Author's collection)

The statue of a miner from the 1940s. (Author's collection)

The statue of a rescue team miner, which faces the former rescue station in Sandy Lane. (Author's collection)

The statue of a miner with the safety gear used before both pits closed. (Author's collection)

Littleton

Littleton Colliery at Huntington was named after Baron Hatherton, who in 1912 had taken the surname Littleton. In 1877 the Cannock & Huntington Colliery Company sank the first No. 1 shaft only to discover 438 feet of water. The workings had flooded by 1880 and, as a result, the company folded in 1884. Later, on 17 February 1899, Lord Hatherton (not to be confused with the Baron!), who owned the estate, began work to sink a new No. 1 shaft to 1,644 feet. Meanwhile, work to recover the original flooded shaft was eventually abandoned on 3 May 1900, leading to a new shaft being sunk to 1,622 feet on 22 November 1902.

In 1912, as part of a national disagreement, Littleton miners went on strike over pay. 6,000 men from Heath Hayes, Cannock, Hednesford and Chadsmoor descended on Littleton, causing thousands of pounds worth of damage. Meanwhile, local police were called to protect the eighty strike-breakers who were underground. To assist the police 500 soldiers of the 1st West Yorkshire Regiment came from Whittingham Barracks from 26 March to 6 April. Twenty-six men were arrested with eight sent to the assizes. One case was discharged and the other seven were found guilty, with three being sentenced to six months of hard labour and the others to four months of hard labour.

Following nationalisation, Littleton Pit was extensively modernised by the National Coal Board. At the end of 1992, the then Conservative Government designated it a 'core pit', offering some protection when other mines were closing. The following year, on 3 December 1993, it closed for good, having been the last working mine of the Cannock Coalfield.

Littleton Colliery in the 1960s.

The wheel from Littleton Colliery is now situated outside the Museum of Cannock Chase. (Author's collection)

The remains of Pillerton Hall, the early seat of the Littleton family before they moved to the now demolished Teddersley Hall, close to Acton Trussel. (Author's collection)

Valley Colliery

Valley Colliery was originally known as 'Pool Pit', after nearby Hednesford Pool which at the time covered Hednesford Park. Opened in 1874, with the first coal being produced a year later, it was the Cannock & Rugeley Colliery Company's second pit. Both shafts were 15 feet in diameter with a depth of 575 feet, while between them a well shaft prevented flooding. Indeed, extracted water was supplied to a number of houses belonging to company officials. Gas, too, was produced until 1882, when the Hednesford Gas Company supplied the mine, probably for lighting.

In 1887 a connection was made with Wimblebury Colliery and all coal was lifted there, although the miners continued to descend via the Valley Colliery's shafts. Then, in 1962, a tunnel of approximately 2 miles was opened between West Cannock No. 5 Colliery and Valley Colliery, resulting in all the workings being incorporated into West Cannock No. 5 Colliery. The Valley Colliery's surface buildings became a training centre before their closure in 1982, while the mine's rescue station, housed in the former baths, remained until 1984. Later, in 1989, Cannock Chase Council took over some of the surface buildings to open the Museum of Cannock Chase.

Cannock Chase Mining Museum, which was originally part of Valley Colliery. (Author's collection)

Walsall Wood Colliery

Work began on sinking two 15-foot-diameter shafts in 1874 on land leased from the Earl of Bradford close to the Wyrley & Essington Canal, where the colliery had its own wharf. There were also links to the London & North Western Railway, the South Staffordshire line and the Walsall Wood branch line. A report by the Walsall Wood Colliery Company on 19 December 1876 explained that No. 1 shaft had required two lengths of tubing with a combined length of 57 yards to progress through the underground water and it was expected that coal would be reached in three weeks' time. When completed the shafts were the deepest in the Cannock Chase Coalfield, at 559 yards for No. 1 Downcast and 581 yards for No. 2 Upcast.

On 29 March 1890 a lamp ignited petrol fumes from an engine that was being tested. Compressed air encouraged the petrol to spray, and one man died while another two suffered burns. Later, on 9 October 1956, a tunnel junction collapsed onto five men,

The Walsall Wood Colliery monument, the tallest pit memorial in South Staffordshire. (Author's collection)

trapping one under a fallen wooden chock and killing four others. One of the rescuers also suffered a broken leg when an unstable rock fell on him. It was the worst mining disaster for twenty-six years.

Until 1954, when it was replaced by an electric fan, Walsall Wood was ventilated by a coal-fired furnace at the bottom of pit – the last mine in the country to have such an arrangement. Coal production eventually finished at Walsall Wood on 30 October 1964 and the mine finally closed in April 1965.

West Cannock No. 1 Plant

The West Cannock Colliery Company Limited was founded in 1869 when they sank their first mine, No. 1 Plant – all their collieries were called Plants – on land leased from the Marquis of Anglesey. It was sited north-west of Hednesford in Pye Green Valley, with both shafts having a diameter of ten feet and No. 1 shaft a winding depth of 301 yards, while No. 2 shaft had a winding depth of 177 yards. Coal production began in 1871. Following nationalisation, the plant was joined underground to No. 3 Plant and its reserves were mined from No. 1. Then, in 1957, West Cannock No. 1 was connected to Littleton Colliery. It closed the following year, in 1958.

West Cannock No. 1 Colliery.

West Cannock No. 2 Plant

Few records survive of West Cannock No. 2 Plant, which was to be found next to the London & North Western Railway line in the Stafford Lane area of Hednesford. The sinking of the two shafts began in 1871 to a depth of 317 metres and the first coal was raised probably in October 1877. The pit has been mistaken for what is known locally as 'West Cannock 2s', which was actually the No. 2 shaft of West Cannock No. 1 Plant. In 1878 a 60-yard roadway was driven through the dividing fault to connect with No. 3 Plant and coal was worked from there. West Cannock No. 2 Plant closed on 4 January 1887.

West Cannock No. 3 Plant

No. 3 Plant sat at the southern end of Pye Green Valley to the south-west of Hednesford, with excavation work beginning there in 1871. Both shafts were 302 yards deep when completed in 1874. Coal was then transported to Hednesford Sidings via the company's railway line, which also served No. 1 Plant, and by tramway to West Cannock Wharf at the Hednesford basin. After nationalisation the mine was connected to No. 1 Plant, and in 1949 it became the third pit to close in the Cannock Chase Coalfield.

West Cannock No. 4 Plant

West Cannock No. 4 Plant, begun in 1869, was around 65 yards from the Company's No. 1 Plant. Indeed, the pit had a single shaft 127 yards deep and a connection with No. 1's Upcast shaft for ventilation. All the surface buildings were equally shared. Later, during the General Strike of 1926, work at the pit stopped and never restarted. No. 4 Plant formally closed in 1928.

West Cannock No. 5 Colliery

West Cannock No. 5 Colliery, formerly No. 5 Plant (known locally as 'Fives' or 'The Tackeroo'), was on the edge of Cannock Chase, overlooking Brindley Heath, and was to the west of the Cannock to Rugeley railway line. It was possibly nicknamed after the Tackeroo military railway, which ran close by as it served the First World War military camps on Cannock Chase. In 1914 the first turf was cut on the Downcast shaft by Stuart Wardle, the former Manager of Nos 1, 3 and 4 Plants, who was by then Company Director and Chairman. Work on the Upcast shaft began in 1916. Progress was hindered due to underground faults and flooding, however, so the mine was not fully operational until 1924/5.

Around 8.30 a.m. on Tuesday 16 May 1933 an explosion in the shallow seam killed six men and two horses, with two other miners suffering burns and thirteen more becoming sick with carbon monoxide poisoning. The disaster was caused when an electric spark made by the ringing of a signal bell ignited gas.

During the ten years following nationalisation in 1947, the pit was renamed Cannock No. 5 Colliery and underwent extensive improvements, including new winding engines and headframes. In all, the cost was £1.75 million. Then, in 1976, the mine suffered more water problems, forcing the abandonment of productive seams. As the remaining reserves became exhausted, the pit was then considered uneconomical. West Cannock No. 5 Colliery finally closed on 17 December 1982.

Wimblebury Colliery

The Cannock & Wimblebury Colliery Company began in 1872, with the first coal being lifted in 1874. The mine lay close to Wimblebury village, not quite a mile from the Valley Colliery Pit. Ideally, it was situated close to the Hednesford canal basin, with a railway siding connecting it to the Littleworth branch line. In November 1880 the company fell into voluntary liquidation, and a second company – the New Cannock & Wimblebury Colliery Company Limited – purchased the pit. Their ownership was short-lived, however, as on 13 July 1887 they too went into liquidation. That year Wimblebury Colliery had its third owners – the Cannock & Rugeley Colliery Company – who purchased the mine for £2,500. The transport links proved an attractive prospect and by 1889 an underground road connected the pit to the nearby Valley Colliery, and thereafter all the coal from both pits was wound at Wimblebury.

On Tuesday 14 June 1927 the newly installed winding engine at Wimblebury was on its second run when an overspeed landing occurred, crashing the cage into the bottom of the shaft. Of the twenty men in the cage, fourteen were injured, five of whom suffered broken legs, and two others died from their injuries. A verdict of accidental death was attributed to the new engine's different controls.

When the National Coal Board took over in 1947, the Wimblebury and Valley site incorporated the reserves of the nearby Cannock Chase Collieries, though production remained limited. The solution was to build a 2-mile connection via the Valley Colliery site to West Cannock No. 5 Pit and transfer all future production there. Wimblebury Colliery closed in 1962 and the shafts were filled in. It is now an industrial site.

Wyrley Grove Colliery

Wyrley Colliery, otherwise known as 'The Grove', lay approximately 2 miles north-west of Brownhills and next to the Cannock extension of the canal, where it had a wharf. Also, a railway led to the Norton branch line. In 1852–3 two shafts were sunk by Messrs William Harrison Limited to a depth of 129 yards. Around 1869 the standard gauge tank engine *Success* was purchased from John Smith of Coven, who supplied engines to several local mines and was the only private engine manufacturer in the area. *Success* was broken up in 1913.

In the late evening of 1 October 1930, an explosion in the shallow seam led to the deaths of fourteen men. Poor ventilation had resulted in the mine being full of carbon monoxide, which, when exposed a naked light, presumed to be a match, ignited. A later search of the coal face proved negative in seeking out a cause. However, six of the dead miners were found to have being carrying contraband cigarettes and live or spent matches. This had been the cause of the explosion. It was the second worst disaster on Cannock Chase Coalfield, beaten only by the tragedy at Pelsall Hall Colliery in November 1872, wherein twenty-two men died.

The village sign for Coven, depicting the colliery railway engines built by John Smith. The building to the rear occupies the site of John Smith's foundry. (Author's collection)

The Brownhills coal miner sculpture by John McKenna, erected in 2006. (Author's collection)

After nationalisation the Wyrley Grove offices became the Group Office for the Cannock Chase area, being also where the wages were processed. Underground production finished in November 1950, but the screening of coal from Wyrley No. 3 Colliery continued until 1963.

Wyrley No. 3 Colliery

When it was sunk in 1896 by William Harrison Limited, the mine was called Brownhills No. 3 Colliery, but was known locally as 'Harrison's' or 'The Sinking'. It lay about halfway between the A5 and the A34 in Wyrley, from where a narrow gauge tramway transported coal to The Grove Colliery for screening – a cheaper alternative to building two plants. However, while the shafts were under construction two men went to inspect the props in the upper part of the shaft. Following on, they asked to be lowered to the water level. The winding engineman never received the signal to stop, however, and the platform carrying both men plunged into the sump water. With their lamps extinguished, one man drowned, while the other managed to hold on to the winding rope and was rescued.

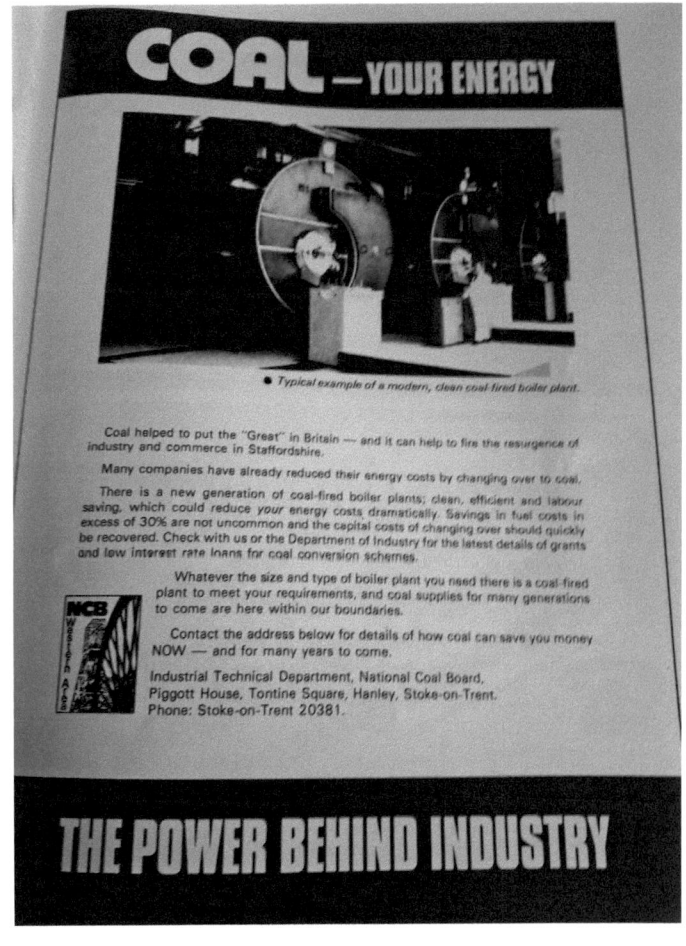

An NCB advert promoting the benefits of coal, c. 1980s. (Author's collection)

Brownhills No. 3 later became an unusual music venue in January 1943, when the BBC broadcast a concert from underground with a violinist, pianist and tenor who had worked in the Welsh pits. Then, in January 1947, the National Coal Board renamed the pit Wyrley No. 3 Colliery and invested in making it more efficient. It was first connected underground with The Grove and then in 1963 with Mid Cannock Colliery. Wyrley No. 3 Colliery finally closed in June 1967.

> ## Coal Industry Estates can offer you
> # PRIME INDUSTRIAL LAND FOR DEVELOPMENT
> ### in various parts of STAFFORDSHIRE
>
> **Berry Hill Estate, Stoke-on-Trent**
> 18 acres in attractive new industrial area with good access to all parts of Potteries.
>
> **Hilton Main, nr. Wolverhampton**
> Up to 18 acres available together with approx. 30,000 sq. ft. in small industrial units. Direct access onto M54.
>
> **Mossfield Industrial Estate, Longton**
> 22 acres available for development in choice area with good links to East Midlands and Potteries.
>
> **Cannock Wood, nr. Cannock**
> 13 acres available together with approx. 9,000 sq. ft. of small refurbished industrial units. Easy reach of A5 and industrial centres.
>
> **Sneyd/Norton Industrial Estates, Stoke-on-Trent**
> Twin estates serving North Staffordshire, limited land available for development in sought-after area.
>
> FOR FURTHER INFORMATION CONTACT: **COAL INDUSTRY ESTATES LTD.,** 72 Leek Road, Stoke-on-Trent. Tel: (0782) 48201

Advert for the sale of former colliery land, *c.* 1980s. (Author's collection)